CRC Series on
COMPUTER-AIDED ENGINEERING

Editor-in-Chief: *Hojjat Adeli*

Neurocomputing for Design Automation
Hojjat Adeli and Hyo Seon Park

High-Performance Computing in Structural Engineering
Hojjat Adeli and Roesdiman Soegiarso

Distributed Computer-Aided Engineering for Analysis, Design, and Visualization
Hojjat Adeli and Sanjay Kumar

HIGH-PERFORMANCE COMPUTING in STRUCTURAL ENGINEERING

Hojjat Adeli
The Ohio State University, U.S.A.

Roesdiman Soegiarso
University of Tarumanagara, Indonesia

CRC Press
Boca Raton   London   New York   Washington, D.C.

Library of Congress Cataloging-in-Publication Data

Adeli, Hojjat, 1950–
    High-performance computing in structural engineering / Hojjat
Adeli, Roesdiman Soegiarso.
        p. cm. -- (CRC series on computer-aided engineering)
    Includes bibliographical references and index.
    ISBN 0-8493-2091-7 (alk. paper)
    1. Structural engineering--Data processing. 2. High performance
computers. 3. Computer-aided engineering. I. Soegiarso,
Roesdiman. II. Title. III. Series.
TA641.A14 1998
624.1′0285—dc21   98-41242
                                                                                                                CIP

    This book contains information obtained from authentic and highly regarded sources. Reprinted material is quoted with permission, and sources are indicated. A wide variety of references are listed. Reasonable efforts have been made to publish reliable data and information, but the author and the publisher cannot assume responsibility for the validity of all materials or for the consequences of their use.

    Neither this book nor any part may be reproduced or transmitted in any form or by any means, electronic or mechanical, including photocopying, microfilming, and recording, or by any information storage or retrieval system, without prior permission in writing from the publisher.

    The consent of CRC Press LLC does not extend to copying for general distribution, for promotion, for creating new works, or for resale. Specific permission must be obtained in writing from CRC Press LLC for such copying.

    Direct all inquiries to CRC Press LLC, 2000 Corporate Blvd., N.W., Boca Raton, Florida 33431.

    Trademark Notice: Product or corporate names may be trademarks or registered trademarks, and are only used for identification and explanation, without intent to infringe.

© 1999 by CRC Press LLC

No claim to original U.S. Government works
International Standard Book Number 0-8493-2091-7
Library of Congress Card Number 98-41242
Printed in the United States of America 1 2 3 4 5 6 7 8 9 0
Printed on acid-free paper

Preface

High-performance multiprocessor computers provide new and interesting opportunities to solve large-scale structural engineering problems. The research challenge is to develop new computational models and algorithms to exploit the unique architecture of these machines. This book is about high-performance computing in structural engineering on supercomputers with vectorization and parallel processing capabilities. Robust parallel-algorithms are presented for analysis and optimization of large structures. The focus of the book is optimization of large structures subjected to actual complicated, implicit, and discontinuous constraints of commonly used design codes. In order to make the book as self-contained as possible, the background and the prerequisite materials are also presented succinctly. The algorithms have been applied to minimum weight design of large steel space trusses and moment-resisting frames with or without bracings consisting of discrete standard shapes and their performance is evaluated. It is shown that for the largest structure with more than nine thousand members a speedup of 17.8 is achieved due to vectorization only. When vectorization is combined with parallel processing, a combined speedup of 99.2 is achieved. This means an efficiency of nearly one hundred compared to the non-vectorized sequential algorithm and code.

Hojjat Adeli and Roesdiman Soegiarso
July 1998

About the authors

Hojjat Adeli received his Ph.D. from Stanford University in 1976. He is currently Professor of Civil and Environmental Engineering and Geodetic Science, Director of the Knowledge Engineering Lab, and a member of the Center for Cognitive Science at The Ohio State University. A contributor to 47 different research journals, he has authored over 300 research and scientific publications in diverse engineering and computer science disciplines. Professor Adeli has authored/co-authored seven pioneering books including *Parallel Processing in Structural Engineering*, Elsevier, 1993; *Machine Learning-Neural Networks, Genetic Algorithms, and Fuzzy Systems*, John Wiley, 1995; *Neurocomputing for Design Automation*, CRC Press, 1998; and *Distributed Computed-Aided Engineering*, CRC Press, 1998. He has also edited 12 books including *Knowledge Engineering – Volume One, Fundamentals,* and *Knowledge Engineering – Volume Two, Applications*, MacGraw-Hill, 1990; *Parallel Processing in Computational Mechanics*, Marcel Dekker, 1992; *Suppercomputing in Engineering Analysis*, Marcel Dekker, 1992; and *Advances in Design Optimization*, Chapman and Hall, 1994. Professor Adeli is the Founder and Editor-in-Chief of the prestigious research journals *Computer-Aided Civil and Infrastructure Engineering* which he founded in 1986 and *Integrated Computer-Aided Engineering* which he founded in 1993. He is the recipient of numerous academic, research, and leadership awards, honors, and recognition. Most recently, he received The Ohio State University Distinguished Scholar Award in 1998. He is listed in 25 Who's Whos and archival biographical listings such as *Two Thousand Notable Americans, Five Hundred Leaders of*

Influence and *Two Thousand Outstanding People of the Twentieth Century*. He has been an organizer or member of advisory board of over 135 national and international conferences and a contributor to 104 conferences held in 35 different countries. He was a Keynote/Plenary Lecturer at 24 international computing conferences held in 20 different countries. Professor Adeli's research has been recognized and sponsored by government funding agencies such as the *National Science Foundation, Federal Highway Administration*, and *U.S. Air Forces Flight Dynamics Lab*, professional organizations such as the *American Iron and Steel Institute* (AISI), the *American Institute of Steel Construction* (AISC); state agencies such as the *Ohio Department of Transportation* and the *Ohio Department of Development*; and private industry such as *Cray Research Inc. and Bethlehem Steel Corporation*. He is a Fellow of the *World Literary Academy* and *American Society of Civil Engineers*.

Roesdiman Soegiarso received his M.S. and Ph.D. degrees in Civil Engineering from The Ohio State University in 1989 and 1994, respectively. He is currently an Assistant Professor of Civil Engineering at the University of Tarumanagara in Jakarta, Indonesia. Dr. Soegiarso is the co-author of seven research articles in the areas of computer-aided design, optimization, and parallel processing.

Acknowledgment

The work presented in this book was partially sponsored by the *National Science Foundation, American Iron and Steel Institute*, and *American Institute of Steel Construction* under grants to the senior author. Supercomputing time on the Cray YMP machines was provided by *The Ohio Supercomputer Center*. Parts of the research presented in this book were published in several articles in *Microcomputers in Civil Engineering* (published by Blackwell Publishers), *Journal of Aerospace Engineering* and *Journal of Structural Engineering* (published by American Society of Civil Engineers), and *Engineering Journal* (published by the American Institute of Steel Construction), as noted in the list of references.

Dedicated to

**Nahid, Anahita, Amir Kevin, Mona
and Cyrus Dean Adeli**

and

Jane, Alex and Astrid Soegiarso

CONTENTS

1. Introduction 1

2. Stiffness Method 5
   2.1 Introduction 5
   2.2 Plane truss 6
      2.2.1 Member stiffness matrix 6
      2.2.2 Coordinate transformation 8
      2.2.3 Transformation of displacements 9
      2.2.4 Transformation of forces 10
      2.2.5 Structure stiffness matrix 12
      2.2.6 Partitioning of the equilibrium equation 13
   2.3 Space trusses 14
      2.3.1 Transformation of displacements 14
   2.4 Plane frames 17
      2.4.1 Member stiffness matrix 17
      2.4.2 Transformation of displacements 24
      2.4.3 Transformation of forces 25
   2.5 Space frames 28
      2.5.1 Member stiffness matrix 28
      2.5.2 Transformation of displacements 38
      2.5.3 Transformation of forces 46

3. Solution of Simultaneous Linear Equations 53
   3.1 Introduction 53
   3.2 LU decomposition 54
   3.3 Cholesky decomposition 57
   3.4 Indirect methods 60
   3.5 Conjugate gradient direction method 62
   3.6 Preconditioned conjugate gradient method 64

  3.7 Global stiffness matrix 66

4. Vectorization Techniques 73
  4.1 Vector processing 73
  4.2 Vectorization techniques... 79

5. Parallel-Vector Algorithms for Analysis of Large
  Structures 85
  5.1 Introduction... 85
  5.2 Concurrent processing 86
  5.3 Concurrent evaluation and assembly of the structure
     stiffness matrix.. 87
  5.4 Concurrent Cholesky and LU decomposition methods 89
  5.5 Concurrent preconditioned conjugate gradient method 97
  5.6 Measuring the performance of parallel-vector
     algorithms 97
  5.7 Application 101
  5.8. Performance results 106

6. Impact of Vectorization on Large-Scale Structural
  Optimization 111
  6.1 Introduction... 111
  6.2 Optimality criteria approach 112
  6.3 Sensitivity analysis 113
  6.4 Recurrence relations for truss structures.................... 117
  6.5 Vectorized optimization algorithm 121
  6.6 Application 130
  6.7 Summary of results and conclusion 135

7. Optimization of Large Steel Structures Using Standard
  Cross Sections.. 139
  7.1 Introduction... 139
  7.2 An optimality criteria approach 140
  7.3 Mapping to standard cross sections 148

    7.4 Application 160
    7.5 Final Comments... 176

8. Parallel-Vector Algorithm for Optimization of Large Structures 179
    8.1 Introduction.. 179
    8.2 An optimality criteria approach 180
    8.3 Parallel-vector algorithm 185
    8.4 Application 202
    8.5 Performance evaluation 214

9. Optimum Load and Resistance Factor Design of Large Steel Space-Frame Structures 223
    9.1 Introduction.. 223
    9.2 Displacement and stress constraints 224
    9.3 Algorithm for structural optimization........................ 226
    9.4 Application 235
    9.5 Comparison of designs based on ASD and LRFD specifications 239
References 241
Subject Index 247

CHAPTER 1
Introduction

1.1 INTRODUCTION

High-performance computing has opened new frontiers to solve large-scale structural engineering problems (Adeli et al., 1994). In their pioneering work and the first book on this subject, Adeli and Kamal (1993) presented parallel algorithms for the analysis and optimization of axial-load truss and moment-resisting frame structures. The parallel algorithms were developed on a prototypical parallel machine, the Encore Multimax (Encore, 1985, 1988). The algorithms were applied to two-dimensional plane frame structures. In this extension of that seminal work, the authors present parallel-vector algorithms developed on a widely used supercomputer, Cray YMP8/864 machine, and apply them to large realistic three-dimensional structures. We employ both vectorization and parallel processing capabilities of the high-performance machines.

How do we design a large structure with a few thousand members? The answer is definitely not unique. A rational design should be based on some logical criterion such as minimizing the weight or cost of the structure. For large structures mathematical optimization algorithms can potentially provide an opportunity to achieve substantial savings in materials and costs.

Optimization of large three-dimensional structures subjected to the actual discontinuous, implicit, and complicated constraints of the commonly used design specifications such as the American Institute of Steel Construction (AISC) Allowable Stress Design (ASD) specifications requires substantial computer processing time (Adeli, 1994; Adeli and Saleh, 1997). With the availability of multiprocessor supercomputers optimization of very large structures consisting of thousands of members is possible (Adeli and Kamal, 1992a&b). Most of the computer processing time in structural optimization algorithms, such as the optimality criteria approach, is spent on iterative arithmetic operations (DO loops). Thus, computer processing time can be reduced by broadening the vectorizable portion of the code. The speedup of an optimization algorithm can be improved significantly through adroit and judicious use of vectorization techniques. Additional speedup can be achieved through the use of parallel processing.

In order to make the book as self-contained as possible, we present an introduction to the stiffness method of structural analysis in Chapter 2 followed by the solution of the linear simultaneous equations in Chapter 3. Chapter 4 presents various approaches to vectorization. Seven methods are described for enhancing vectorization. In Chapter 5 parallel-vector algorithms are presented for analysis of large space structures.

Parallelization and vectorization of assembling the stiffness matrix in the structural analysis problem and solution of the resulting system of linear equations are studied. Both direct and indirect methods are investigated to solve the resulting system of linear equations. The impact of vectorization on the performance of structural optimization algorithms for structures of various size is investigated in Chapter 6.

In the practical design of steel structures only a finite number of shapes are available, such as those given in the AISC manuals (AISC, 1989, 1994). In Chapter 7, a multi-constraint optimality criteria discrete optimization algorithm is presented for minimum weight design of large steel structures subjected to stress, displacement, and buckling constraints specified by the AISC ASD or LRFD specifications. An efficient integer mapping strategy is presented for mapping the computed cross-sectional areas to the available, standard, wide flange (W) shapes.

In Chapter 8 a robust and efficient parallel-vector multi-constraint discrete optimization algorithm is presented for optimization of large space structures subjected to the constraints of the AISC ASD specifications. The types of structures considered are space axial load structures, moment-resisting frames, and moment-resisting frames with diagonal bracings. Optimization of steel moment-resisting space frame structures with or without cross-bracings subjected to AISC LRFD specifications is presented in Chapter 9.

CHAPTER 2
Stiffness Method

2.1 INTRODUCTION

There are basically two different approaches for the analysis of structures subjected to loadings: stiffness or displacement method and flexibility or force method. Since the advent of computers the stiffness method has been used as the preferred approach for its adaptability to programming on digital computers. The stiffness method of structural analysis is presented in this chapter. In this approach, the primary unknowns are the nodal displacements which are found by solving a set of linear simultaneous equations. After finding the nodal displacements, the member displacements and forces are found. We limit our attention to framed structures including plane and space trusses and plane and space frames.

6 Stiffness Method

We make the following assumptions throughout this chapter and the rest of the book:
1. The material is homogeneous and isotropic.
2. Displacements are small and the change of geometry is negligible.

2.2 PLANE TRUSSES

2.2.1 MEMBER STIFFNESS MATRIX

We first determine the member stiffness matrix for a prismatic truss member. A truss member with end nodes 1 and 2 and length L_m is shown in Figure 2.1. The global coordinate axes are x and y and the local coordinate axes are x' and y'. The cross-sectional area of the member is A and the modulus of elasticity is E. By definition, stiffness coefficients are forces created by unit displacements. Figure 2.1b shows the stiffness coefficients k_{11} and k_{21} due to a unit displacement at end 1 of the member. Similarly, Figure 2.1c shows stiffness coefficients k_{12} and k_{22} due to a unit displacement at end 2 of the member. Using Hooke's law it can readily be shown that

$$k_{11} = \frac{AE}{L_m} \tag{2.1}$$

$$k_{21} = k_{12} = -\frac{AE}{L_m} \tag{2.2}$$

$$k_{22} = \frac{AE}{L_m} \tag{2.3}$$

In matrix form we can write the 4x4 stiffness matrix for member m in the member local coordinates as

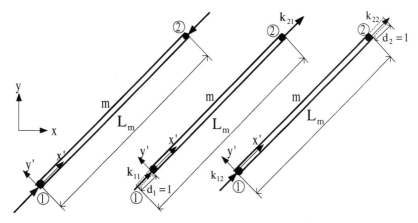

a) Member m   b) End 1 is displaced by 1   c) End 2 is displaced by 1

Figure 2.1: The stiffness coefficients of member m in local coordinates

$$\mathbf{k}_m^l = \frac{AE}{L_m}\begin{bmatrix} 1 & -1 \\ -1 & 1 \end{bmatrix} \qquad (2.4)$$

When the external loads f_1 and f_2 are acting simultaneously on member m, as shown in Figure 2.2, the displacements at node 1, d_1, and node 2, d_2, can be related to the forces as follows:

$$\begin{Bmatrix} f_1 \\ f_2 \end{Bmatrix} = \frac{AE}{L_m}\begin{bmatrix} 1 & -1 \\ -1 & 1 \end{bmatrix}\begin{Bmatrix} d_1 \\ d_2 \end{Bmatrix} \qquad (2.5)$$

or in matrix form,

$$\mathbf{f}_m = \mathbf{k}_m^l \mathbf{d}_m \qquad (2.6)$$

8 Stiffness Method

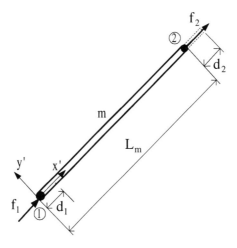

Figure 2.2: Forces and displacements on member m in local coordinates

2.2.2 COORDINATE TRANSFORMATION

Before we can assemble the local stiffness matrices of the members of a truss into the global structure stiffness matrix, each member stiffness matrix must be transformed from its local coordinate axes x' and y' to the global coordinate axes x and y. Suppose the local axis x' of a member m with ends i and j makes an angle of α with the global axis x (Figure 2.3). Denoting the global x and y coordinates of end i by x_i and y_i and end j by x_j and y_j, the direction cosines of the axis of the member m are

$$\lambda_x = \cos\alpha = \frac{x_j - x_i}{\sqrt{(x_j - x_i)^2 + (y_j - y_i)^2}} \qquad (2.7)$$

$$\lambda_y = \sin\alpha = \frac{y_j - y_i}{\sqrt{(x_j - x_i)^2 + (y_j - y_i)^2}} \qquad (2.8)$$

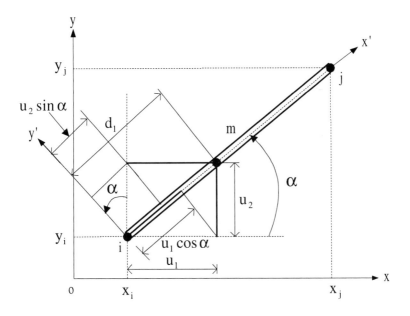

Figure 2.3: Coordinate transformation of displacements

2.2.3 TRANSFORMATION OF DISPLACEMENTS

We denote the global displacements of node i in the x and y directions by u_1 and u_2 and of node j by u_3 and u_4. We also denote the displacements of nodes i and j in the direction of the local x' axis by d_1 and d_2, respectively (Figure 2.3). The global and local nodal displacements can be related by

$$d_1 = u_1 \cos \alpha + u_2 \sin \alpha \qquad (2.9)$$

$$d_2 = u_3 \cos \alpha + u_4 \sin \alpha \qquad (2.10)$$

In matrix form Eqs. (2.9) and (2.10) can be written as follows:

$$\mathbf{d}_m = \mathbf{T}_m \mathbf{u}_m \qquad (2.11)$$

where

$$\mathbf{d}_m = \begin{Bmatrix} d_1 \\ d_2 \end{Bmatrix} \tag{2.12}$$

$$\mathbf{T}_m = \begin{bmatrix} \lambda_x & \lambda_y & 0 & 0 \\ 0 & 0 & \lambda_x & \lambda_y \end{bmatrix} \tag{2.13}$$

$$\mathbf{u}_m = \begin{Bmatrix} u_1 \\ u_2 \\ u_3 \\ u_4 \end{Bmatrix} \tag{2.14}$$

2.2.4 TRANSFORMATION OF FORCES

Let f_1 be the force at end i of the member in the direction of the local x' axis and F_1 and F_2 be the forces at the same end i in the direction of global axes x and y (Figure 2.4). We can write,

$$F_1 = f_1 \cos \alpha \tag{2.15}$$

$$F_2 = f_1 \sin \alpha \tag{2.16}$$

Similarly, the global components of forces at end j, F_3 and F_4, are related to the force f_2 in the direction of the local x' axis by

$$F_3 = f_2 \cos \alpha \tag{2.17}$$

$$F_4 = f_2 \sin \alpha \tag{2.18}$$

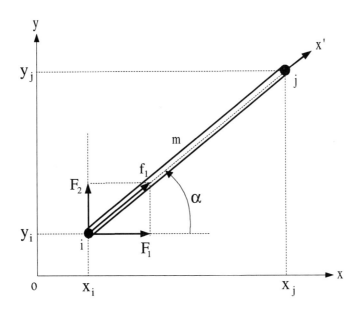

Figure 2.4: Transformation of forces

In matrix form, Eqs. (2.15) through (2.18) can be expressed as

$$\begin{Bmatrix} F_1 \\ F_2 \\ F_3 \\ F_4 \end{Bmatrix} = \begin{bmatrix} \lambda_x & 0 \\ \lambda_y & 0 \\ 0 & \lambda_x \\ 0 & \lambda_y \end{bmatrix} \begin{Bmatrix} f_1 \\ f_2 \end{Bmatrix} \qquad (2.19)$$

or

$$\mathbf{F}_m = \mathbf{T}_m^T \mathbf{f}_m \qquad (2.20)$$

where $\mathbf{F}_m$ is the column vector of member forces in the global coordinates.

Substituting Eq. (2.11) into Eq. (2.6) yields

$$\mathbf{f}_m = \mathbf{k}_m^1 \mathbf{T}_m \mathbf{u}_m \tag{2.21}$$

Substituting for $\mathbf{f}_m$ from Eq. (2.21) into Eq. (2.20) we obtain

$$\mathbf{F}_m = \mathbf{T}_m^T \mathbf{k}_m^1 \mathbf{T}_m \mathbf{u}_m \tag{2.22}$$

or

$$\mathbf{F}_m = \mathbf{k}_m \mathbf{u}_m \tag{2.23}$$

where $\mathbf{k}_m$ is the member stiffness matrix in global coordinates.

$$\mathbf{k}_m = \mathbf{T}_m^T \mathbf{k}_m^1 \mathbf{T}_m \tag{2.24}$$

$$\mathbf{k}_m = \begin{bmatrix} \lambda_x & 0 \\ \lambda_y & 0 \\ 0 & \lambda_x \\ 0 & \lambda_y \end{bmatrix} \frac{AE}{L_m} \begin{bmatrix} 1 & -1 \\ -1 & 1 \end{bmatrix} \begin{bmatrix} \lambda_x & \lambda_y & 0 & 0 \\ 0 & 0 & \lambda_x & \lambda_y \end{bmatrix} \tag{2.25}$$

$$[\mathbf{k}_m] = \frac{AE}{L_m} \begin{bmatrix} l^2 & lm & -l^2 & -lm \\ lm & m^2 & -lm & -m^2 \\ -l^2 & -lm & l^2 & lm \\ -lm & -m^2 & lm & m^2 \end{bmatrix} \tag{2.26}$$

where $l = \lambda_x$ and $m = \lambda_y$

2.2.5 STRUCTURE STIFFNESS MATRIX

By assembling all the member matrices, we can obtain the global structure stiffness matrix:

$$\mathbf{K} = \sum_{m=1}^{M} \mathbf{k}_m \tag{2.27}$$

where M is the total number of members in the structure. The equilibrium equation in the global coordinates becomes

F = K u (2.28)

where **F** is the $2N_n \times 1$ column matrix of the external nodal forces and **u** is the $2N_n \times 1$ column matrix of nodal displacements including the restrained displacements. N_n is the number of nodes in the structure. The dimensions of the structure stiffness matrix **K** are $2N_n \times 2N_n$.

2.2.6 PARTITIONING OF THE EQUILIBRIUM EQUATION

The equilibrium equation (2.28) can be partitioned in the following form:

$$\begin{Bmatrix} \mathbf{F}_n \\ \mathbf{F}_s \end{Bmatrix} = \begin{bmatrix} \mathbf{k}_{11} & \mathbf{k}_{12} \\ \mathbf{k}_{21} & \mathbf{k}_{22} \end{bmatrix} \begin{Bmatrix} \mathbf{u}_n \\ \mathbf{u}_s \end{Bmatrix} \quad (2.29)$$

or in expanded form

$$\mathbf{F}_n = \mathbf{k}_{11}\mathbf{u}_n + \mathbf{k}_{12}\mathbf{u}_s \quad (2.30)$$

$$\mathbf{F}_s = \mathbf{k}_{21}\mathbf{u}_n + \mathbf{k}_{22}\mathbf{u}_s \quad (2.31)$$

where the subscripts n and s denote the nodal displacement degrees of freedom and nodal support-restrained degrees of freedom, respectively. Assuming no displacements at restrained support degrees of freedom, $\mathbf{u}_s = 0$, and Eqs. (2.30) and (2.31) are simplified to

$$\mathbf{F}_n = \mathbf{k}_{11}\mathbf{u}_n \quad (2.32)$$

$$\mathbf{F}_s = \mathbf{k}_{21}\mathbf{u}_n \tag{2.33}$$

Solution of Eq. (2.32) yields the nodal displacements, $\mathbf{u}_n$. Having found the nodal displacements, the support reactions can be found from Eq. (2.33). Then, the nodal displacements of member m in local coordinates are found from Eq. (2.11). Next, member forces in local coordinates can be found from Eq. (2.6).

2.3 SPACE TRUSSES

2.3.1 TRANSFORMATION OF DISPLACEMENTS

We denote the global displacements of node i in the global x, y and z directions by u_1, u_2 and u_3 and of node j by u_4, u_5 and u_6 (Figure 2.5). We also denote the displacements of nodes i and j in the direction of the local x' axis by d_1 and d_2, respectively (Figure 2.6). The global and local nodal displacements can be related by

$$d_1 = u_1 \cos\alpha + u_2 \cos\beta + u_3 \cos\gamma \tag{2.34}$$

$$d_2 = u_4 \cos\alpha + u_5 \cos\beta + u_6 \cos\gamma \tag{2.35}$$

where $\lambda_x = \cos\alpha$, $\lambda_y = \cos\beta$, and $\lambda_z = \cos\gamma$ are the direction cosines determined as follows:

$$\lambda_x = \cos\alpha = \frac{x_j - x_i}{L_m} \tag{2.36}$$

$$\lambda_y = \cos\beta = \frac{y_j - y_i}{L_m} \tag{2.37}$$

$$\lambda_z = \cos\gamma = \frac{z_j - z_i}{L_m} \tag{2.38}$$

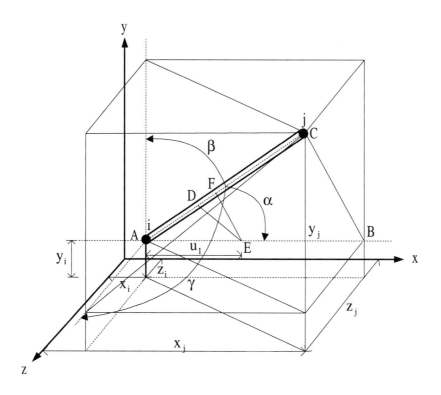

Figure 2.5: Coordinates transformation

In matrix form, Eqs. (2.34) and (2.35) can be written as follows:

$$\mathbf{d}_m = \mathbf{T}_m \mathbf{u}_m \qquad (2.39)$$

where

$$\mathbf{d}_m = \begin{Bmatrix} d_1 \\ d_2 \end{Bmatrix} \qquad (2.40)$$

16 Stiffness Method

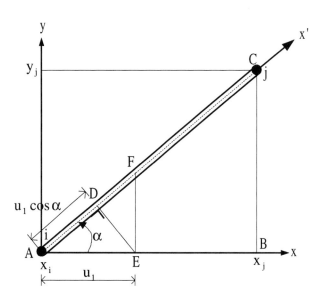

Figure 2.6: The relationship between local and global displacements

$$\mathbf{T}_m = \begin{bmatrix} \lambda_x & \lambda_y & \lambda_z & 0 & 0 & 0 \\ 0 & 0 & 0 & \lambda_x & \lambda_y & \lambda_z \end{bmatrix} \quad (2.41)$$

$$\mathbf{u}_m = \begin{Bmatrix} u_1 \\ u_2 \\ u_3 \\ u_4 \\ u_5 \\ u_6 \end{Bmatrix} \quad (2.42)$$

Similar to the plane truss, the member stiffness matrix for a space truss can be written as follows:

$$\mathbf{k}_m = \mathbf{T}_m^T \mathbf{k}_m^l \mathbf{T}_m \tag{2.43}$$

$$\mathbf{k}_m = \begin{bmatrix} \lambda_x & 0 \\ \lambda_y & 0 \\ \lambda_z & 0 \\ 0 & \lambda_x \\ 0 & \lambda_y \\ 0 & \lambda_z \end{bmatrix} \frac{AE}{L_m} \begin{bmatrix} 1 & -1 \\ -1 & 1 \end{bmatrix} \begin{bmatrix} \lambda_x & \lambda_y & \lambda_z & 0 & 0 & 0 \\ 0 & 0 & 0 & \lambda_x & \lambda_y & \lambda_z \end{bmatrix} \tag{2.44}$$

$$[\mathbf{k}_m] = \frac{AE}{L_m} \begin{bmatrix} l^2 & lm & ln & -l^2 & -lm & -ln \\ lm & m^2 & mn & -lm & -m^2 & -mn \\ ln & mn & n^2 & -ln & -mn & -n^2 \\ -l^2 & -lm & -ln & l^2 & lm & ln \\ -lm & -m^2 & -mn & lm & m^2 & mn \\ -ln & -mn & -n^2 & ln & mn & n^2 \end{bmatrix} \tag{2.45}$$

where $l = \lambda_x$, $m = \lambda_y$ and $n = \lambda_z$.

2.4 PLANE FRAMES

2.4.1 MEMBER STIFFNESS MATRIX

A typical member m in a plane frame is shown in Figure 2.7. The local member axes are x' and y'. The global coordinate axes are x and y. The cross-sectional area of the member is A and its moment of inertia is I. The member end forces and their corresponding displacements are noted in Figure 2.7. At end i, there are axial force f_1, shear force, f_2, and

18 Stiffness Method

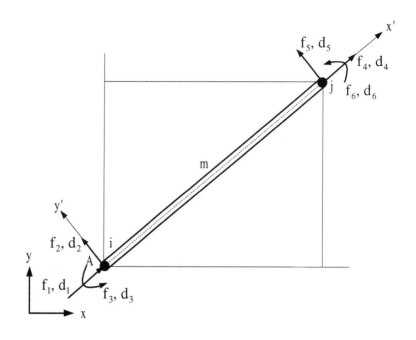

Figure 2.7: Forces and displacements of member m in local coordinates

bending moment f_3 with corresponding displacements d_1, d_2, and d_3, respectively. Similarly, at end j, there are axial force f_4, shear force, f_5, and bending moment f_6 with corresponding displacements d_4, d_5, and d_6, respectively. In Figure 2.7, all the forces and displacements are drawn in their positive directions. Stiffness coefficients are, by definition, forces created by unit displacements and rotations.

a) Stiffness coefficients due to a unit axial displacement

Figure 2.8 shows the stiffness coefficients k_{11} and k_{41} due to a unit axial displacement at end i in the direction of the local x′ axis. Using Hooke's law it can easily be shown

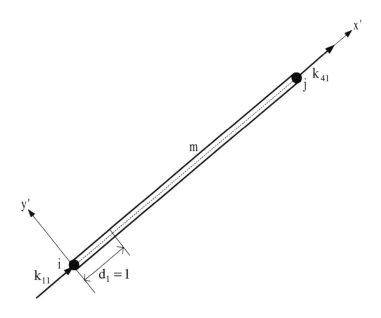

Figure 2.8: Stiffness coefficients k_{11} and k_{41} due to a unit displacement at end i in the direction of local x' axis

$$k_{11} = \frac{EA}{L_m} \qquad (2.46)$$

$$k_{41} = -\frac{EA}{L_m} \qquad (2.47)$$

Similarly, the stiffness coefficients k_{44} and k_{14} due to a unit axial displacement at end j of the member in the direction of the local x' axis are obtained as

$$k_{44} = \frac{EA}{L_m} \qquad (2.48)$$

$$k_{14} = -\frac{EA}{L_m} \qquad (2.49)$$

b) Stiffness coefficients due to a unit transverse displacement

Figure 2.9 shows the stiffness coefficients k_{22}, k_{32}, k_{52}, and k_{62} due to a unit transverse displacement at end i in the direction of the local y' axis. Using the simple Bernoulli-Euler beam theory we can write

$$k_{22} = \frac{12EI}{L_m^3} \tag{2.50}$$

$$k_{32} = \frac{6EI}{L_m^2} \tag{2.51}$$

$$k_{52} = -\frac{12EI}{L_m^3} \tag{2.52}$$

$$k_{62} = \frac{6EI}{L_m^2} \tag{2.53}$$

Similarly, the stiffness coefficients k_{25}, k_{35}, k_{55}, and k_{65} due to a unit transverse displacement at end j of the member in the direction of local y' axis are obtained as

$$k_{25} = -\frac{12EI}{L_m^3} \tag{2.54}$$

$$k_{35} = -\frac{6EI}{L_m^2} \tag{2.55}$$

$$k_{55} = \frac{12EI}{L_m^3} \tag{2.56}$$

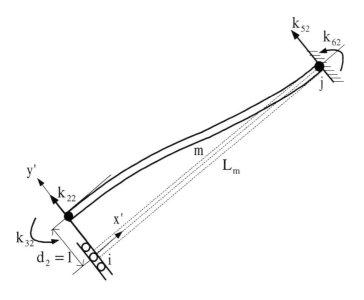

Figure 2.9: Stiffness coefficients k_{22}, k_{32}, k_{52}, and k_{62} due to a unit displacement in the direction of local y' axis

$$k_{65} = -\frac{6EI}{L_m^2} \qquad (2.57)$$

c) Stiffness coefficients due to a unit rotation

Figure 2.10 shows the stiffness coefficients k_{23}, k_{33}, k_{53}, and k_{63} due to a unit rotation at end i of the member. Again, using the simple Bernoulli-Euler beam theory we can write

$$k_{23} = \frac{6EI}{L_m^2} \qquad (2.58)$$

$$k_{33} = \frac{4EI}{L_m} \qquad (2.59)$$

22 Stiffness Method

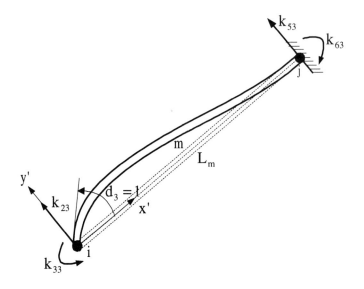

Figure 2.10: Stiffness coefficients k_{23}, k_{33}, k_{53}, and k_{63} due to a unit rotation at end i

$$k_{53} = -\frac{6EI}{L_m^2} \tag{2.60}$$

$$k_{63} = \frac{2EI}{L_m} \tag{2.61}$$

Similarly, the stiffness coefficients k_{26}, k_{36}, k_{56}, and k_{66} due to a unit rotation at end j of the member are obtained as

$$k_{26} = \frac{6EI}{L_m^2} \tag{2.62}$$

$$k_{36} = \frac{2EI}{L_m} \tag{2.63}$$

$$k_{56} = -\frac{6EI}{L_m^2} \tag{2.64}$$

$$k_{66} = \frac{4EI}{L_m} \tag{2.65}$$

$$\begin{Bmatrix} f_1 \\ f_2 \\ f_3 \\ f_4 \\ f_5 \\ f_6 \end{Bmatrix} = \begin{bmatrix} \frac{EA}{L_m} & 0 & 0 & -\frac{EA}{L_m} & 0 & 0 \\ 0 & \frac{12EI}{L_m^3} & \frac{6EI}{L_m^2} & 0 & -\frac{12EI}{L_m^3} & \frac{6EI}{L_m^2} \\ 0 & \frac{6EI}{L_m^2} & \frac{4EI}{L_m} & 0 & -\frac{6EI}{L_m^2} & \frac{2EI}{L_m} \\ -\frac{EA}{L_m} & 0 & 0 & \frac{EA}{L_m} & 0 & 0 \\ 0 & -\frac{12EI}{L_m^3} & -\frac{6EI}{L_m^2} & 0 & \frac{12EI}{L_m^3} & -\frac{6EI}{L_m^2} \\ 0 & \frac{6EI}{L_m^2} & \frac{2EI}{L_m} & 0 & -\frac{6EI}{L_m^2} & \frac{4EI}{L_m} \end{bmatrix} \begin{Bmatrix} d_1 \\ d_2 \\ d_3 \\ d_4 \\ d_5 \\ d_6 \end{Bmatrix} \tag{2.66}$$

Equations (2.46) to (2.65) represent various terms of the member stiffness matrix in the local coordinates. They can be written in the following matrix form:

or

$$\mathbf{f}_m = \mathbf{k}_m^l \mathbf{d}_m \tag{2.67}$$

2.4.2 TRANSFORMATION OF DISPLACEMENTS

We denote the global displacements of node i in the global x and y directions by u_1 and u_2 and of node j by u_4 and u_5 (Figure 2.11). We also denote the displacements of nodes i in the directions of the local x' and y' axis by d_1 and d_2, respectively. The global and local nodal displacements at end i can be related by

$$d_1 = u_1 \cos \alpha + u_2 \sin \alpha \tag{2.68}$$

$$d_2 = -u_1 \sin \alpha + u_2 \cos \alpha \tag{2.69}$$

Since the global z and local z' axes are coincident, the rotation d_3 about the z' axis is the same as rotation u_3 about the z axis.

$$d_3 = u_3 \tag{2.70}$$

Similar to end i, the global and local nodal displacements at end j can be related by

$$d_4 = u_4 \cos \alpha + u_5 \sin \alpha \tag{2.71}$$

$$d_5 = -u_4 \sin \alpha + u_5 \cos \alpha \tag{2.72}$$

$$d_6 = u_6 \tag{2.73}$$

In matrix form, Eqs. (2.68) through (2.73) can be expressed as

$$\begin{Bmatrix} d_1 \\ d_2 \\ d_3 \\ d_4 \\ d_5 \\ d_6 \end{Bmatrix} = \begin{bmatrix} \lambda_x & \lambda_y & 0 & 0 & 0 & 0 \\ -\lambda_y & \lambda_x & 0 & 0 & 0 & 0 \\ 0 & 0 & 1 & 0 & 0 & 0 \\ 0 & 0 & 0 & \lambda_x & \lambda_y & 0 \\ 0 & 0 & 0 & -\lambda_y & \lambda_x & 0 \\ 0 & 0 & 0 & 0 & 0 & 1 \end{bmatrix} \begin{Bmatrix} u_1 \\ u_2 \\ u_3 \\ u_4 \\ u_5 \\ u_6 \end{Bmatrix} \tag{2.74}$$

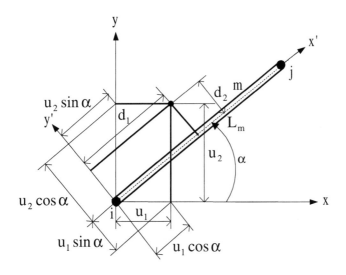

Figure 2.11: The relationship between local and global displacements

or
$$\mathbf{d}_m = \mathbf{T}_m \mathbf{u}_m \qquad (2.75)$$
where $\mathbf{d}_m$ is the 6x1 column matrix of the nodal displacements in the local member coordinates, $\mathbf{T}_m$ is the 6x6 displacement transformation matrix, and $\mathbf{u}_m$ is the 6x1 column matrix of the nodal displacements in the global coordinates.

2.4.3 TRANSFORMATION OF FORCES

Let f_1 and f_2 be the forces at end i of the member m in the direction of the local x' and y' axes, respectively, and let F_1 and F_2 be the forces at the same end i in the direction of the global x and y axes (Figure 2.12). We can write

26 Stiffness Method

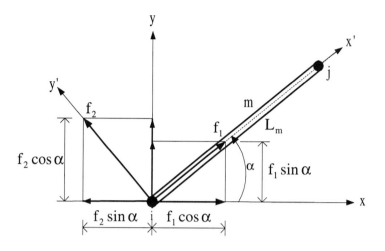

Figure 2.12: Relationship between global and local coordinates

$$F_1 = f_1 \cos \alpha - f_2 \sin \alpha \qquad (2.76)$$

$$F_2 = f_1 \sin \alpha + f_2 \cos \alpha \qquad (2.77)$$

Since the global z and local z' axes coincide, the bending moment at end i about the z axis, F_3, is the same as the bending moment about the z' axis.

$$F_3 = f_3 \qquad (2.78)$$

Similarly, the global components of forces at end j, F_4, F_5 and F_6 are related to the forces f_4, f_5 and f_6 in the directions of the local x' and y' axes by

$$F_4 = f_4 \cos \alpha - f_5 \sin \alpha \qquad (2.79)$$

$$F_5 = f_4 \sin \alpha + f_5 \cos \alpha \qquad (2.80)$$

$$F_6 = f_6 \tag{2.81}$$

In matrix form, Eqs. (2.76) through (2.81) can be written as follows:

$$\begin{Bmatrix} F_1 \\ F_2 \\ F_3 \\ F_4 \\ F_5 \\ F_6 \end{Bmatrix} = \begin{bmatrix} \lambda_x & -\lambda_y & 0 & 0 & 0 & 0 \\ \lambda_y & \lambda_x & 0 & 0 & 0 & 0 \\ 0 & 0 & 1 & 0 & 0 & 0 \\ 0 & 0 & 0 & \lambda_x & -\lambda_y & 0 \\ 0 & 0 & 0 & \lambda_y & \lambda_x & 0 \\ 0 & 0 & 0 & 0 & 0 & 1 \end{bmatrix} \begin{Bmatrix} f_1 \\ f_2 \\ f_3 \\ f_4 \\ f_5 \\ f_6 \end{Bmatrix} \tag{2.82}$$

or

$$\mathbf{F}_m = \mathbf{T}_m^T \mathbf{f}_m \tag{2.83}$$

where $\mathbf{F}_m$ is the 6x1 column matrix of nodal forces in global coordinates and $\mathbf{f}_m$ is the 6x1 column matrix of member nodal forces in the local coordinates.

Substituting Eq. (2.67) into Eq. (2.83) yields

$$\mathbf{F}_m = \mathbf{T}_m^T \mathbf{k}_m^l \mathbf{d}_m \tag{2.84}$$

Substituting Eq. (2.75) into Eq. (2.84), we obtain

$$\mathbf{F}_m = \mathbf{T}_m^T \mathbf{k}_m^l \mathbf{T}_m \mathbf{u}_m \tag{2.85}$$

or

$$\mathbf{F}_m = \mathbf{k}_m \mathbf{u}_m \tag{2.86}$$

where $\mathbf{k}_m$ is the element stiffness matrix in the global coordinates.

$$\mathbf{k}_m = \mathbf{T}_m^T \mathbf{k}_m^l \mathbf{T}_m \qquad (2.87)$$

2.5 SPACE FRAMES

2.5.1 MEMBER STIFFNESS MATRIX

A typical member m in a space frame is shown in Figure 2.13. The local member axes are x', y' and z'. The global coordinate axes are x, y and z. The cross-sectional area of the member is A and its moments of inertia about the y' and z' axes are I_y and I_z, respectively. G is the shear modulus of the elasticity and J is the polar moment of inertia of the cross section. The member forces and their corresponding displacements are noted in Figure 2.13. At end i there are axial force f_1; shear force in the direction of y' axis, f_2; shear force in the direction of z' axis, f_3; with the corresponding displacements d_1, d_2, and d_3, respectively. In addition, there are twisting moment, f_4, and bending moments, f_5 and f_6, with the corresponding rotations d_4 about x' axis, d_5 about y' axis, and d_6 about z' axis. Similarly, at end j there are axial force f_7; shear force in the direction of y' axis, f_8; shear force in the direction of z' axis, f_9; with the corresponding displacements d_7, d_8 and d_9, respectively. In addition, there are twisting moment, f_{10}, and bending moments, f_{11} and f_{12}, with the corresponding rotations d_{10} about x' axis, d_{11} about y' axis, and d_{12} about z' axis.

a) Stiffness coefficients due to a unit axial displacement

Figure 2.14 shows the stiffness coefficients k_{11} and k_{71} due to a unit axial displacement at end i in the direction of local x' axis. Using Hooke's law we can write

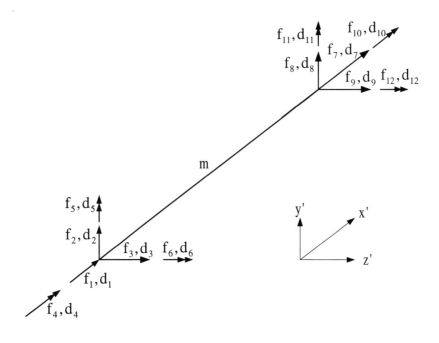

Figure 2.13: Forces and displacements of member m in local coordinates

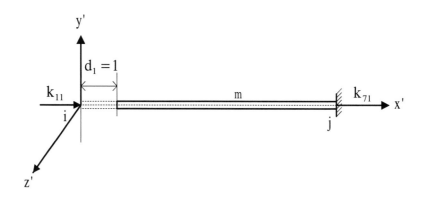

Figure 2.14: Stiffness coefficients k_{11} and k_{71} due to a unit displacement at end i in the direction of the local x' axis

$$k_{11} = \frac{EA}{L_m} \tag{2.88}$$

$$k_{71} = -\frac{EA}{L_m} \tag{2.89}$$

Similarly, the stiffness coefficients k_{77} and k_{17} due to a unit axial displacement at end j of the member in the direction of the local x' axis are obtained as

$$k_{77} = \frac{EA}{L_m} \tag{2.90}$$

$$k_{17} = \frac{EA}{L_m} \tag{2.91}$$

b) Stiffness coefficients due to a unit transverse displacement in the direction of y' axis

Figure 2.15 shows the stiffness coefficients k_{22}, k_{62}, k_{82}, and $k_{12,2}$ due to a unit transverse displacement at end i in the direction of the local y' axis. Using the simple beam theory we can write

$$k_{2,2} = \frac{12EI_z}{L_m^3} \tag{2.92}$$

$$k_{8,2} = -\frac{12EI_z}{L_m^3} \tag{2.93}$$

$$k_{6,2} = \frac{6EI_z}{L_m^2} \tag{2.94}$$

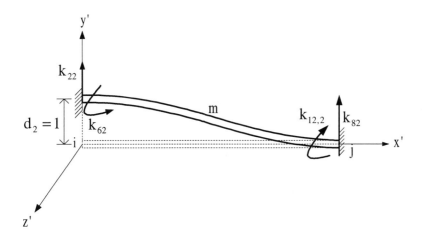

Figure 2.15: Stiffness coefficients k_{22}, k_{62}, k_{82}, and $k_{12,2}$ due to a unit transverse displacement in the direction of the local y' axis

$$k_{12,2} = \frac{6EI_z}{L_m^2} \tag{2.95}$$

Similarly, the stiffness coefficients k_{28}, k_{68}, k_{88}, and $k_{12,8}$ due to a unit transverse displacement at end j in the direction of the local y' axis are obtained as

$$k_{2,8} = -\frac{12EI_z}{L_m^3} \tag{2.96}$$

$$k_{6,8} = -\frac{6EI_z}{L_m^2} \tag{2.97}$$

$$k_{8,8} = \frac{12EI_z}{L_m^3} \tag{2.98}$$

$$k_{12,8} = -\frac{6EI_z}{L_m^2} \tag{2.99}$$

c) Stiffness coefficients due to a unit transverse displacement in the direction of z' axis

Figure 2.16 shows the stiffness coefficients k_{33}, k_{53}, k_{93} and $k_{11,3}$ due to a unit axial displacement at end i in the direction of the local z' axis. Using the simple beam theory we can write

$$k_{3,3} = \frac{12EI_y}{L_m^3} \tag{2.100}$$

$$k_{5,3} = -\frac{6EI_y}{L_m^2} \tag{2.101}$$

$$k_{9,3} = -\frac{12EI_y}{L_m^3} \tag{2.102}$$

$$k_{11,3} = -\frac{6EI_y}{L_m^2} \tag{2.103}$$

Similarly, the stiffness coefficients k_{33}, k_{53}, k_{93}, and $k_{11,3}$ due to a unit transverse displacement at end j in the direction of the local z' axis are obtained as

$$k_{3,9} = -\frac{12EI_y}{L_m^3} \tag{2.104}$$

$$k_{5,9} = \frac{6EI_y}{L_m^2} \tag{2.105}$$

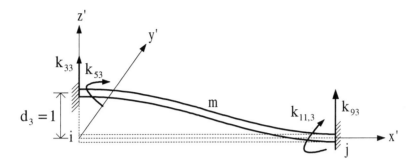

Figure 2.16: Stiffness coefficients k_{33}, k_{53}, k_{93}, and $k_{11,3}$ due to a unit displacement in the direction of the local z' axis

$$k_{9,9} = \frac{12EI_y}{L_m^3} \qquad (2.106)$$

$$k_{11,9} = \frac{6EI_y}{L_m^2} \qquad (2.107)$$

d) Stiffness coefficients due to a unit rotation about the x' axis

Figure 2.17 shows the stiffness coefficients k_{44} and $k_{10,4}$ due to a unit rotation about the local x' axis at end i. They are expressed as

$$k_{4,4} = \frac{GJ}{L_m} \qquad (2.108)$$

$$k_{10,4} = -\frac{GJ}{L_m} \qquad (2.109)$$

34 Stiffness Method

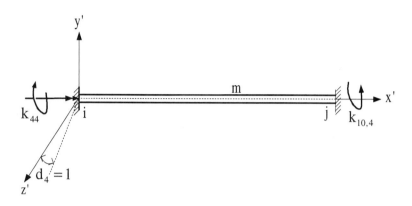

Figure 2.17: Stiffness coefficients k_{44} and $k_{10,4}$ due to a unit rotation about the local x' axis at end i

Similarly, the stiffness coefficients $k_{4,10}$ and $k_{10,10}$ due to a unit rotation about the local x' axis at end j are obtained as

$$k_{4,10} = -\frac{GJ}{L_m} \tag{2.110}$$

$$k_{10,10} = \frac{GJ}{L_m} \tag{2.111}$$

e) Stiffness coefficients due to a unit rotation about the local y' axis

Figure 2.18 shows the stiffness coefficients k_{35}, k_{55}, k_{95}, and $k_{11,5}$ due to a unit rotation at end i about the local y' axis. Using the simple beam theory we can write

$$k_{3,5} = -\frac{6EI_y}{L_m^2} \tag{2.112}$$

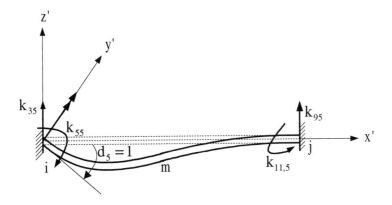

Figure 2.18: Stiffness coefficients k_{35}, k_{55}, k_{95}, and $k_{11,5}$ due to a unit rotation about the local y' axis at end i

$$k_{5,5} = \frac{4EI_y}{L_m} \tag{2.113}$$

$$k_{9,5} = \frac{6EI_y}{L_m^2} \tag{2.114}$$

$$k_{11,5} = \frac{2EI_y}{L_m} \tag{2.115}$$

Similarly, the stiffness coefficients $k_{3,11}, k_{5,11}, k_{9,11}$, and $k_{11,11}$ due to a unit rotation about the local y' axis at end j are obtained as

$$k_{3,11} = -\frac{6EI_y}{L_m^2} \tag{2.116}$$

36 Stiffness Method

$$k_{5,11} = \frac{2EI_y}{L_m} \qquad (2.117)$$

$$k_{9,11} = \frac{6EI_y}{L_m^2} \qquad (2.118)$$

$$k_{11,11} = \frac{4EI_y}{L_m} \qquad (2.119)$$

f) Stiffness coefficients due to a unit rotation about the local z' axis

Figure 2.19 shows the stiffness coefficients k_{26}, k_{66}, k_{86}, and $k_{12,6}$ due to a unit rotation about the local z' axis at end i. Using the simple beam theory we can write

$$k_{2,6} = \frac{6EI_z}{L_m^2} \qquad (2.120)$$

$$k_{6,6} = \frac{4EI_z}{L_m} \qquad (2.121)$$

$$k_{8,6} = -\frac{6EI_z}{L_m^2} \qquad (2.122)$$

$$k_{12,6} = \frac{2EI_z}{L_m} \qquad (2.123)$$

Similarly, the stiffness coefficients $k_{2,12}, k_{6,12}, k_{8,12}$ and $k_{12,12}$ due to a unit rotation about the local z' axis at end j are obtained as

$$k_{2,12} = \frac{6EI_z}{L_m^2} \qquad (2.124)$$

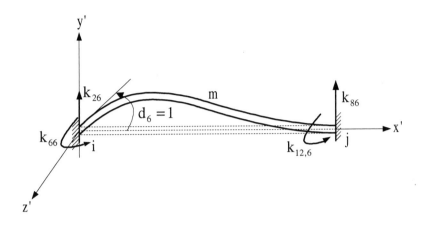

Figure 2.19: Stiffness coefficients k_{26}, k_{66}, k_{86}, and $k_{12,6}$ due to a unit rotation at end i about the local z' axis

$$k_{6,12} = \frac{2EI_z}{L_m} \qquad (2.125)$$

$$k_{9,12} = -\frac{6EI_z}{L_m^2} \qquad (2.126)$$

$$k_{12,12} = \frac{4EI_z}{L_m} \qquad (2.127)$$

Equations (2.88) to (2.127) represent various terms of the member stiffness matrix in the local coordinates. They can be written in the following matrix form:

38 Stiffness Method

$$\mathbf{k}'_m = \begin{bmatrix}
\frac{EA}{L_m} & 0 & 0 & 0 & 0 & 0 & -\frac{EA}{L_m} & 0 & 0 & 0 & 0 & 0 \\
0 & \frac{12EI_z}{L_m^3} & 0 & 0 & 0 & \frac{6EI_z}{L_m^2} & 0 & -\frac{12EI_z}{L_m^3} & 0 & 0 & 0 & \frac{6EI_z}{L_m^2} \\
0 & 0 & \frac{12EI_y}{L_m^3} & 0 & -\frac{6EI_y}{L_m^2} & 0 & 0 & 0 & -\frac{12EI_y}{L_m^3} & 0 & -\frac{6EI_y}{L_m^2} & 0 \\
0 & 0 & 0 & \frac{GJ}{L_m} & 0 & 0 & 0 & 0 & 0 & -\frac{GJ}{L_m} & 0 & 0 \\
0 & 0 & -\frac{6EI_y}{L_m^2} & 0 & \frac{4EI_y}{L_m} & 0 & 0 & 0 & \frac{6EI_y}{L_m^2} & 0 & \frac{2EI_y}{L_m} & 0 \\
0 & \frac{6EI_z}{L_m^2} & 0 & 0 & 0 & \frac{4EI_z}{L_m} & 0 & -\frac{6EI_z}{L_m^2} & 0 & 0 & 0 & \frac{2EI_z}{L_m} \\
-\frac{EA}{L_m} & 0 & 0 & 0 & 0 & 0 & \frac{EA}{L_m} & 0 & 0 & 0 & 0 & 0 \\
0 & -\frac{12EI_z}{L_m^3} & 0 & 0 & 0 & -\frac{6EI_z}{L_m^2} & 0 & \frac{12EI_z}{L_m^3} & 0 & 0 & 0 & -\frac{6EI_z}{L_m^2} \\
0 & 0 & -\frac{12EI_y}{L_m^3} & 0 & \frac{6EI_y}{L_m^2} & 0 & 0 & 0 & \frac{12EI_y}{L_m^3} & 0 & \frac{6EI_y}{L_m^2} & 0 \\
0 & 0 & 0 & -\frac{GJ}{L_m} & 0 & 0 & 0 & 0 & 0 & \frac{GJ}{L_m} & 0 & 0 \\
0 & 0 & -\frac{6EI_y}{L_m^2} & 0 & \frac{2EI_y}{L_m} & 0 & 0 & 0 & \frac{6EI_y}{L_m^2} & 0 & \frac{4EI_y}{L_m} & 0 \\
0 & \frac{6EI_z}{L_m^2} & 0 & 0 & 0 & \frac{2EI_z}{L_m} & 0 & -\frac{6EI_z}{L_m^2} & 0 & 0 & 0 & \frac{4EI_z}{L_m}
\end{bmatrix} \quad (2.128)$$

2.5.2 TRANSFORMATION OF DISPLACEMENTS

For plane frame structures, the transformation matrix is described by a single rotation α about the z axis perpendicular to the xy plane, as shown in Figure 2.11. To establish the displacement transformation matrix for space frame structures we have to consider three rotations about three different coordinate axes. Suppose the coordinate axes xyz are transformed to the coordinate axes x'y'z' through rotations α about x axis, β about y axis, and γ about z axis. In Figure 2.20, first the xyz coordinate axes are transformed to x*y*z* axes through rotation β about y axis. Next, the x*y*z* axes are transformed to the x"y"z" axes through rotation γ about the z* axis. Finally, the x"y"z" axes are transformed to the x'y'z' axes through rotation α about the x" axis. In Section 2.5.1 we defined the 12 components of the member end displacements, d_1 to d_{12}, in the local coordinate axes (Figure 2.13). Similarly, we define the corresponding displacements in the direction of the global axes: u_1, u_2, and u_3 are the three linear displacements in the directions of the global axes and u_4, u_6 and u_6 are the three rotations about the global axes for end i of the member. Displacements u_7 to u_{12} are defined similarly for end j.

a) Transformation of displacements due to rotation β about the y axis

Consider a member m lying in the xz plane as shown in Figure 2.21. The nodal displacements of end i of the member in the directions of x*, y*, and z* axes, d_1^*, d_2^*, and d_3^*, respectively, are related to u_1, u_2, and u_3 in the directions of global axes by

$$d_1^* = u_1 \cos\beta + u_3 \sin\beta \tag{2.129}$$

$$d_2^* = u_2 \tag{2.130}$$

$$d_3^* = -u_1 \sin\beta + u_3 \cos\beta \tag{2.131}$$

In matrix form, Eqs. (2.128) through (2.130) can be expressed as

$$\begin{Bmatrix} d_1^* \\ d_2^* \\ d_3^* \end{Bmatrix} = \begin{bmatrix} \cos\beta & 0 & \sin\beta \\ 0 & 1 & 0 \\ -\sin\beta & 0 & \cos\beta \end{bmatrix} \begin{Bmatrix} u_1 \\ u_2 \\ u_3 \end{Bmatrix} \tag{2.132}$$

40 Stiffness Method

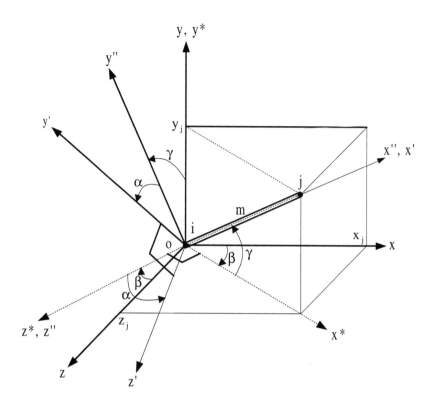

Figure 2.20: Transformation from global to local coordinate axes

As such, the three linear displacement components are related through the transformation matrix:

$$\mathbf{T}_y = \begin{bmatrix} \cos\beta & 0 & \sin\beta \\ 0 & 1 & 0 \\ -\sin\beta & 0 & \cos\beta \end{bmatrix} \qquad (2.133)$$

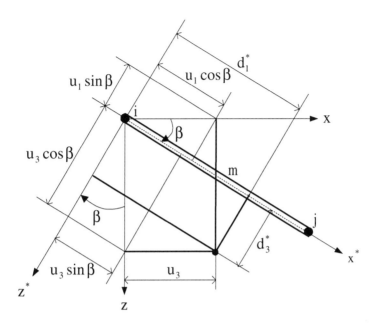

Figure 2.21: The relationship between displacements in the directions of x and z and x* and z* axes due to rotation β about the y axis

b) Transformation of displacements due to rotation γ about the z* axis

Consider a member m lying in the x*y* plane as shown in Figure 2.22. The nodal displacements of end i of the member in the directions of x", y", and z", d_1'', d_2'', and d_3'', respectively, are related to displacements d_1^*, d_2^*, and d_3^* in the directions of the x*y*z* axes by

$$d_1'' = d_1^* \cos\gamma + d_2^* \sin\gamma \tag{2.134}$$

$$d_2'' = -d_1^* \sin\gamma + d_2^* \cos\gamma \tag{2.135}$$

42 Stiffness Method

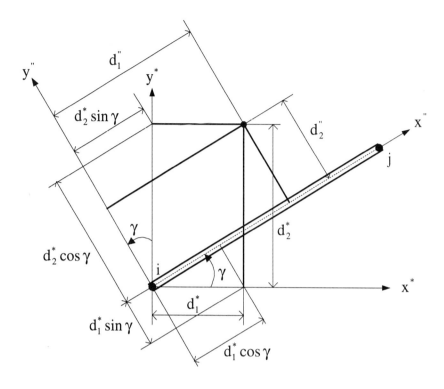

Figure 2.22: The relationship between displacements in the directions of x* and y* and x" and y" axes due to rotation γ about the z* axis

$$d_3'' = d_3^* \tag{2.136}$$

In matrix form, Eqs. (2.134) – (2.136) can be written as

$$\begin{Bmatrix} d_1'' \\ d_2'' \\ d_3'' \end{Bmatrix} = \begin{bmatrix} \cos\gamma & \sin\gamma & 0 \\ -\sin\gamma & \cos\gamma & 0 \\ 0 & 0 & 1 \end{bmatrix} \begin{Bmatrix} d_1^* \\ d_2^* \\ d_3^* \end{Bmatrix} \tag{2.137}$$

Consequently, the three linear displacement components are related through the 3x3 transformation matrix

$$T_z = \begin{bmatrix} \cos\gamma & \sin\gamma & 0 \\ -\sin\gamma & \cos\gamma & 0 \\ 0 & 0 & 1 \end{bmatrix} \quad (2.138)$$

c) Transformation of displacements due to rotation α about the x" axis

Consider a member m lying in the y" z" plane as shown in Figure 2.23. The nodal displacements of end i of the member in the directions x', y', and z' axes, d_1', d_2', and d_3', respectively, are related to displacements d_1'', d_2'', and d_3'' in the directions of the x" y" z" axes by

$$d_1' = d_1'' \quad (2.139)$$

$$d_2' = d_2'' \cos\alpha + d_3'' \sin\alpha \quad (2.140)$$

$$d_3' = -d_2'' \sin\alpha + d_3'' \cos\alpha \quad (2.141)$$

In matrix form, these equations can be written as

$$\begin{Bmatrix} d_1' \\ d_2' \\ d_3' \end{Bmatrix} = \begin{bmatrix} 1 & 0 & 0 \\ 0 & \cos\alpha & \sin\alpha \\ 0 & -\sin\alpha & \cos\alpha \end{bmatrix} \begin{Bmatrix} d_1'' \\ d_2'' \\ d_3'' \end{Bmatrix} \quad (2.142)$$

As such, the three linear displacement components are related through the 3x3 transformation matrix

$$T_x = \begin{bmatrix} 1 & 0 & 0 \\ 0 & \cos\alpha & \sin\alpha \\ 0 & -\sin\alpha & \cos\alpha \end{bmatrix} \quad (2.143)$$

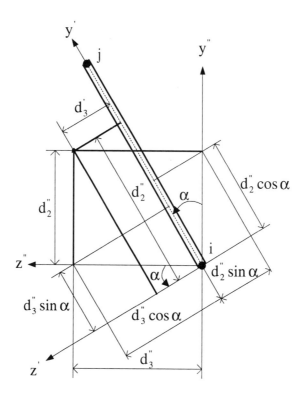

Figure 2.23: The relationship between displacements in the directions of the y" and z" and y' and z' axes due to rotation α about the x" axes

d) Transformation of displacements due to rotations about three axes

Combining the three transformation equations (2.131), (2.137), and (2.142), we obtain the total transformation matrix

$$\begin{Bmatrix} d_1' \\ d_2' \\ d_3' \end{Bmatrix} = \begin{bmatrix} 1 & 0 & 0 \\ 0 & \cos\alpha & \sin\alpha \\ 0 & -\sin\alpha & \cos\alpha \end{bmatrix} \begin{bmatrix} \cos\gamma & \sin\gamma & 0 \\ -\sin\gamma & \cos\gamma & 0 \\ 0 & 0 & 1 \end{bmatrix} \begin{bmatrix} \cos\beta & 0 & \sin\beta \\ 0 & 1 & 0 \\ -\sin\beta & 0 & \cos\beta \end{bmatrix} \begin{Bmatrix} u_1 \\ u_2 \\ u_3 \end{Bmatrix} \quad (2.144)$$

or

$$\begin{Bmatrix} d_1' \\ d_2' \\ d_3' \end{Bmatrix} = \begin{bmatrix} \cos\beta\cos\gamma & \sin\gamma & \sin\beta\cos\gamma \\ -\cos\alpha\cos\beta\sin\gamma - \sin\alpha\sin\beta & \cos\alpha\cos\gamma & -\cos\alpha\sin\beta\sin\gamma + \sin\alpha\cos\beta \\ \sin\alpha\cos\beta\sin\gamma - \cos\alpha\sin\beta & -\sin\alpha\cos\gamma & \sin\alpha\sin\beta\sin\gamma + \cos\alpha\cos\beta \end{bmatrix} \begin{Bmatrix} u_1 \\ u_2 \\ u_3 \end{Bmatrix} \quad (2.145)$$

Equation (2.145) represents the transformation matrix for linear displacements at end i of the member due to rotations of axes. Similar relations hold for transformations of linear displacements at end j of the member as well as rotations at each end of the member. Consequently, we obtain the following transformation relation in matrix form between the 12x1 column matrices of the local and global displacements $\mathbf{d}_m$ and $\mathbf{u}_m$:

$$\mathbf{d}_m = \begin{bmatrix} \mathbf{T}_d & 0 & 0 & 0 \\ 0 & \mathbf{T}_d & 0 & 0 \\ 0 & 0 & \mathbf{T}_d & 0 \\ 0 & 0 & 0 & \mathbf{T}_d \end{bmatrix} \mathbf{u}_m \quad (2.146)$$

where

$$\{\mathbf{T}_d\} = \begin{bmatrix} \cos\beta\cos\gamma & \sin\gamma & \sin\beta\cos\gamma \\ -\cos\alpha\cos\beta\sin\gamma - \sin\alpha\sin\beta & \cos\alpha\cos\gamma & -\cos\alpha\sin\beta\sin\gamma + \sin\alpha\cos\beta \\ \sin\alpha\cos\beta\sin\gamma - \cos\alpha\sin\beta & -\sin\alpha\cos\gamma & \sin\alpha\sin\beta\sin\gamma + \cos\alpha\cos\beta \end{bmatrix}$$

$$(2.147)$$

Equation (3.146) can be written as follows:
$$\mathbf{d}_m = \mathbf{T}_m \mathbf{u}_m \quad (2.148)$$

where $\mathbf{T}_m$ is the 12x12 transformation matrix.

2.5.3 TRANSFORMATION OF FORCES

Similar to the transformation of displacements, we compute the transformation matrices for rotations about various coordinate axes.

a) Transformation of forces due to rotation β about y axis.

Let f_1^*, f_2^* and f_3^* be the forces at end i of member m in the directions of x*, y* and z* axes, respectively, and let F_1, F_2 and F_3 be the forces at the same end i in the directions of x, y and z axes, respectively (Figure 2.24). We can write the following relationship

$$F_1 = f_1^* \cos\beta - f_3^* \sin\beta \qquad (2.149)$$

$$F_2 = f_2^* \qquad (2.150)$$

$$F_3 = f_1^* \sin\beta + f_3^* \cos\beta \qquad (2.151)$$

Equations. (2.149) through (2.151) can be written in matrix form as follows:

$$\mathbf{T}_y^T = \begin{bmatrix} \cos\beta & 0 & -\sin\beta \\ 0 & 1 & 0 \\ \sin\beta & 0 & \cos\beta \end{bmatrix} \qquad (2.152)$$

or

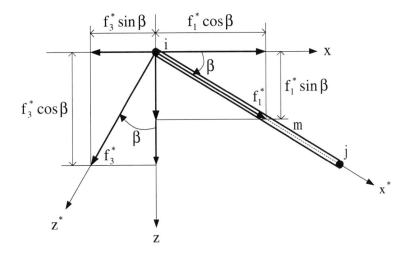

Figure 2.24: Relationship between local and global component of forces

$$\mathbf{T}_y^T = \begin{bmatrix} \cos\beta & 0 & -\sin\beta \\ 0 & 1 & 0 \\ \sin\beta & 0 & \cos\beta \end{bmatrix} \quad (2.153)$$

where $\mathbf{T}_y^T$ is the transpose of transformation matrix $\mathbf{T}_y$ given by Eq. (2.132).

b) Transformation of forces due to rotation γ about the z* axis

Let f_1'', f_2'' and f_3'' be the forces at end i of member m in the directions of the x'', y'' and z'' axes, respectively, and let f_1^*, f_2^* and f_3^* be the forces at the same end i in the directions of the x*, y* and z* axes, respectively (Figure 2.25). We can write the following relationship

48 Stiffness Method

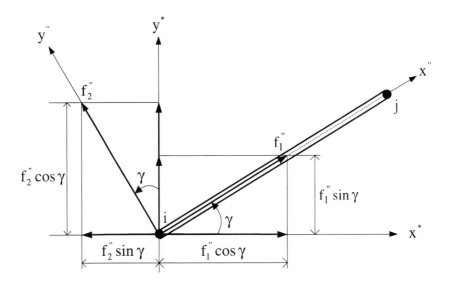

Figure 2.25: Relationship between local and global component of forces

$$f_1^* = f_1^{''} \cos\gamma - f_2^{''} \sin\gamma \tag{2.154}$$

$$f_2^* = f_1^{''} \sin\gamma + f_2^{''} \cos\gamma \tag{2.155}$$

$$f_3^* = f_3^{''} \tag{2.156}$$

Equations. (2.154) through (2.156) can be written in matrix form as follows:

$$\begin{Bmatrix} f_1^* \\ f_2^* \\ f_3^* \end{Bmatrix} = \begin{bmatrix} \cos\gamma & -\sin\gamma & 0 \\ \sin\gamma & \cos\gamma & 0 \\ 0 & 0 & 1 \end{bmatrix} \begin{Bmatrix} f_1^{''} \\ f_2^{''} \\ f_3^{''} \end{Bmatrix} \tag{2.157}$$

or

$$\mathbf{T}_z^T = \begin{bmatrix} \cos\gamma & -\sin\gamma & 0 \\ \sin\gamma & \cos\gamma & 0 \\ 0 & 0 & 1 \end{bmatrix} \quad (2.158)$$

where $\mathbf{T}_z^T$ is the transpose of transformation matrix $\mathbf{T}_z$ given by Eq. (2.138).

b) Transformation of forces due to rotation α about the x" axis

Let f_1', f_2' and f_3' be the forces at end i of member m in the directions of the x', y' and z' axes, respectively, and let f_1'', f_2'' and f_3'' be the forces at the same end i in the direction of the x", y" and z" axes, respectively (Figure 2.26). We can write the following relationship

$$f_1'' = f_1' \quad (2.159)$$

$$f_2'' = f_2' \cos\alpha - f_3' \sin\alpha \quad (2.160)$$

$$f_3'' = f_2' \sin\alpha + f_3' \cos\alpha \quad (2.161)$$

Equations (2.159) through (2.161) can be written in matrix form as follows:

$$\begin{Bmatrix} f_1'' \\ f_2'' \\ f_3'' \end{Bmatrix} = \begin{bmatrix} 1 & 0 & 0 \\ 0 & \cos\alpha & -\sin\alpha \\ 0 & \sin\alpha & \cos\alpha \end{bmatrix} \begin{Bmatrix} f_1' \\ f_2' \\ f_3' \end{Bmatrix} \quad (2.162)$$

or

50 Stiffness Method

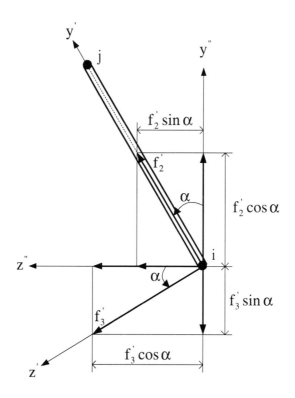

Figure 2.26: Relationship between local and global component of forces

$$\mathbf{T}_x^T = \begin{bmatrix} 1 & 0 & 0 \\ 0 & \cos\alpha & -\sin\alpha \\ 0 & \sin\alpha & \cos\alpha \end{bmatrix} \qquad (2.163)$$

where $\mathbf{T}_x^T$ is the transpose of transformation matrix $\mathbf{T}_x$ given by Eq. (3.143)

c) Transformation of forces due to rotations about three axes

Combining the three transformation equations, (2.153), (2.158), (2.163), we obtain the total transformation matrix

$$\begin{Bmatrix} F_1 \\ F_2 \\ F_3 \end{Bmatrix} = \begin{bmatrix} \cos\beta & 0 & -\sin\beta \\ 0 & 1 & 0 \\ \sin\beta & 0 & \cos\beta \end{bmatrix} \begin{bmatrix} \cos\gamma & -\sin\gamma & 0 \\ \sin\gamma & \cos\gamma & 0 \\ 0 & 0 & 1 \end{bmatrix} \begin{bmatrix} 1 & 0 & 0 \\ 0 & \cos\alpha & -\sin\alpha \\ 0 & \sin\alpha & \cos\alpha \end{bmatrix} \begin{Bmatrix} f_1' \\ f_2' \\ f_3' \end{Bmatrix} \quad (2.164)$$

or

$$\begin{Bmatrix} F_1 \\ F_2 \\ F_3 \end{Bmatrix} = \begin{bmatrix} \cos\beta\cos\gamma & -\cos\alpha\cos\beta\sin\gamma - \sin\alpha\sin\beta & \sin\alpha\cos\beta\sin\gamma - \cos\alpha\sin\beta \\ \sin\gamma & \cos\alpha\cos\gamma & -\sin\alpha\cos\gamma \\ \sin\beta\cos\gamma & -\cos\alpha\sin\beta\sin\gamma + \sin\alpha\cos\beta & \sin\alpha\sin\beta\sin\gamma + \cos\alpha\cos\beta \end{bmatrix}$$
$$\begin{Bmatrix} f_1' \\ f_2' \\ f_3' \end{Bmatrix} \quad (2.165)$$

Equation (2.165) represents the transformation matrix for linear forces at end i of the member due to rotations of axes. Similar relations hold for transformations of linear forces at end j of the member as well as rotations at each end of the member. Consequently, we obtain the following transformation relation in matrix form between the 12x1 column matrices of the local and global forces $\mathbf{F}_m$ and $\mathbf{f}_m'$:

$$\mathbf{F}_m = \begin{bmatrix} \mathbf{T}_d^T & 0 & 0 & 0 \\ 0 & \mathbf{T}_d^T & 0 & 0 \\ 0 & 0 & \mathbf{T}_d^T & 0 \\ 0 & 0 & 0 & \mathbf{T}_d^T \end{bmatrix} \mathbf{f}_m' \quad (2.166)$$

where

$$\mathbf{T}_d^T = \begin{bmatrix} \cos\beta\cos\gamma & -\cos\alpha\cos\beta\sin\gamma - \sin\alpha\sin\beta & \sin\alpha\cos\beta\sin\gamma - \cos\alpha\sin\beta \\ \sin\gamma & \cos\alpha\cos\gamma & -\sin\alpha\cos\gamma \\ \sin\beta\cos\gamma & -\cos\alpha\sin\beta\sin\gamma + \sin\alpha\cos\beta & \sin\alpha\sin\beta\sin\gamma + \cos\alpha\cos\beta \end{bmatrix}$$

(2.167)

Equation (2.166) can be written as follows:

$$\mathbf{F}_m = \mathbf{T}_m^T \mathbf{f}_m' \tag{2.168}$$

where $\mathbf{T}_m^T$ is the 12x12 forces transformation matrix.

CHAPTER 3
Solution of Simultaneous Linear Equations

3.1. INTRODUCTION

Solution of the system of simultaneous linear equations is the most time-consuming part of the structural analysis problem. For large structures, more than 90% of the CPU time may be spent in the solution of the simultaneous linear equations. For the structural optimization problem, this part is even more taxing because of the iterative nature of the nonlinear programming problem and the need for repeated solution of the set of linear equations. In this chapter, three popular methods for the solution of simultaneous linear equations are presented. They are the LU decomposition, Cholesky decomposition, and the preconditioned conjugate gradient (PCG) methods. The first two methods are called direct methods where the solution is

54 Solution of Simultaneous Linear Equations

obtained directly without any iterations. The third method is an indirect or iterative method.

3.2 LU DECOMPOSITION

Suppose the solution of the set of simultaneous linear equations

$$\mathbf{K}\mathbf{u} = \mathbf{F} \tag{3.1}$$

is desired. In the **LU** decomposition approach, matrix **K** is decomposed into the product of a lower triangular matrix **L** with the diagonal terms equal to one and an upper triangular matrix **U** in the following form:

$$\mathbf{K} = \mathbf{L}\mathbf{U} \tag{3.2}$$

$$\mathbf{L} = \begin{bmatrix} 1 & 0 & \cdots & 0 \\ L_{21} & 1 & \cdots & 0 \\ \cdots & \cdots & \cdots & \cdots \\ L_{n1} & L_{n2} & \cdots & 1 \end{bmatrix} \tag{3.3}$$

$$\mathbf{U} = \begin{bmatrix} U_{11} & U_{12} & \cdots & U_{1n} \\ 0 & U_{22} & \cdots & U_{2n} \\ \cdots & \cdots & \cdots & \cdots \\ 0 & 0 & \cdots & U_{nn} \end{bmatrix} \tag{3.4}$$

Substituting Eq. (3.2) into Eq. (3.1) yields

$$\mathbf{L}\mathbf{U}\mathbf{u} = \mathbf{F} \tag{3.5}$$

Letting

$$\mathbf{U}\mathbf{u} = \mathbf{v} \tag{3.6}$$

Equation (3.5) will be transformed to

$$\mathbf{L}\mathbf{v} = \mathbf{F} \tag{3.7}$$

The solution of linear Eq. (3.1) now consists of three steps. In the first step, **K** is decomposed into the product of **L** and **U**. The second step is the forward solution of Eq. (3.7) to find **v**. The final step is the backward solution of Eq. (3.6) for **u**.

Equation (3.2) in expanded form becomes

$$\begin{bmatrix} K_{11} & K_{12} & \cdots & K_{n1} \\ K_{21} & K_{22} & \cdots & K_{2n} \\ \cdots & \cdots & \cdots & \cdots \\ K_{n1} & K_{n2} & \cdots & K_{nn} \end{bmatrix} = \begin{bmatrix} U_{11} & U_{12} & \cdots & U_{1n} \\ L_{21}U_{11} & L_{21}U_{12}+U_{22} & \cdots & L_{21}U_{1n}+U_{2n} \\ \cdots & \cdots & \cdots & \cdots \\ L_{n1}U_{11} & L_{n1}U_{12}+L_{n2}U_{22} & \cdots & L_{n1}U_{1n}+L_{n2}U_{2n}+..+U_{nn} \end{bmatrix} \tag{3.8}$$

The elements of the first column of **K** are

$$\begin{cases} K_{11} = U_{11} \\ K_{21} = L_{21}U_{11} \quad \rightarrow \quad L_{21} = K_{21}/U_{11} \\ \cdots \\ K_{n1} = L_{n1}U_{11} \quad \rightarrow \quad L_{n1} = K_{n1}/U_{11} \end{cases} \tag{3.9}$$

The elements of the second column of **K** are

$$\begin{cases} K_{12} = U_{12} \\ K_{22} = L_{21}U_{12} + U_{22} \quad \rightarrow \quad U_{22} = K_{22} - L_{21}U_{12} \\ \cdots \\ K_{n2} = L_{n1}U_{12} + L_{n2}U_{22} \quad \rightarrow \quad L_{n2} = \dfrac{K_{n2} - L_{n1}U_{12}}{U_{22}} \end{cases} \tag{3.10}$$

The elements of the nth column of **K** are

$$\begin{cases} K_{1n} = U_{1n} \\ K_{2n} = L_{21}U_{12} + U_{2n} \\ \dots \\ K_{nn} = L_{n1}U_{1n} + L_{n2}U_{2n} + \dots + U_{nn} \end{cases} \begin{array}{l} \rightarrow U_{2n} = K_{2n} - L_{21}U_{12} \\ \\ \rightarrow U_{nn} = K_{nn} - L_{n1}U_{1n} - \\ \quad L_{n2}U_{2n} - \dots - L_{n(n-1)}U_{(n-1)n} \end{array} \quad (3.11)$$

An algorithm for decomposing the matrix **K** is presented in Table 3.1. Note that in this algorithm, lower and upper triangular matrices **L** and **U** are stored in the same locations of the matrix **K** in order to minimize the required storage. The algorithms for forward and backward solutions are given in Tables 3.2 and 3.3, respectively.

Table 3.1 Algorithm for decomposition of K into lower and upper triangular matrices (LU decomposition)

Loop 1 $i = 1$ until $n-1$
   Loop 2 $j = i+1$ until n
      $K_{ji} = K_{ji} / K_{ii}$
   Endloop 2
   Loop 3 $k = i+1$ until n
      Loop 4 $m = i+1$ until n
         $K_{mk} = K_{mk} - K_{mi}K_{ik}$
      Endloop 4
   Endloop 3
Endloop 1

Table 3.2 Algorithm for forward solution of the LU decomposition method

Loop 1 i = 2 until n
  Loop 2 j = 1 until i − 1
    $F_i = F_i - K_{ij}F_j$
  Endloop 2
Endloop 1

Table 3.3 Algorithm for backward solution of the LU decomposition method

Loop 1 i = n until 1
  $F_i = F_i / K_{ii}$
  j = i − 1
  Loop 2 k = i + 1 until n
    $F_j = F_j - K_{ik}F_k$
  Endloop 2
  $F_j = F_j / K_{jj}$
Endloop 1

3.3 CHOLESKY DECOMPOSITION

The Cholesky factorization method requires that matrix **K** in Eq. (3.1) be positive definite in addition to being symmetric. In this approach matrix **K** is decomposed into the product of a lower triangular matrix **L** and its transpose $\mathbf{L}^T$ which is an upper triangular matrix.

$$\mathbf{K} = \mathbf{LL}^T \tag{3.12}$$

This equation in expanded form is written as

$$\begin{bmatrix} K_{11} & K_{12} & \cdots & K_{n1} \\ K_{21} & K_{22} & \cdots & 0 \\ \cdots & \cdots & \cdots & \cdots \\ K_{n1} & K_{n2} & \cdots & K_{nn} \end{bmatrix} = \begin{bmatrix} L_{11} & 0 & \cdots & 0 \\ L_{21} & L_{22} & \cdots & 0 \\ \cdots & \cdots & \cdots & \cdots \\ L_{n1} & L_{n2} & \cdots & L_{nn} \end{bmatrix} \begin{bmatrix} L_{11} & L_{12} & \cdots & L_{1n} \\ 0 & L_{22} & \cdots & L_{2n} \\ \cdots & \cdots & \cdots & \cdots \\ 0 & 0 & \cdots & L_{nn} \end{bmatrix} \quad (3.13)$$

The elements of the first column of the lower triangular matrix **L** are found from the following sequence of equations:

$$\begin{cases} K_{11} = L_{11}L_{11} & \rightarrow \quad L_{11} = \sqrt{K_{11}} \\ K_{21} = L_{21}L_{11} & \rightarrow \quad L_{21} = \dfrac{K_{21}}{L_{11}} \\ \cdots \\ K_{n1} = L_{n1}L_{11} & \rightarrow \quad L_{n1} = \dfrac{K_{n1}}{L_{11}} \end{cases} \quad (3.14)$$

The nonzero elements of the second column of **L** (below the diagonal term) are found from the following sequence of equations:

$$\begin{cases} K_{22} = L_{21}L_{21} + L_{22}L_{22} & \rightarrow \quad L_{22} = \sqrt{K_{22} - L_{21}L_{21}} \\ K_{32} = L_{31}L_{21} + L_{32}L_{22} & \rightarrow \quad L_{32} = \dfrac{K_{32} - L_{31}L_{21}}{L_{22}} \\ \cdots \\ K_{n2} = L_{n1}U_{21} + L_{n2}L_{22} & \rightarrow \quad L_{n2} = \dfrac{K_{n2} - L_{n1}L_{21}}{L_{22}} \end{cases} \quad (3.15)$$

Similar equations can be written for the remaining columns. There is only one nonzero term in the last column, the diagonal term, which is found as follows:

$$K_{nn} = L_{nn}L_{nn} \rightarrow L_{nn} = \sqrt{K_{nn}} \qquad (3.16)$$

Equations (3.14) through (3.16) can be summarized as follows:

$$\begin{cases} L_{11} = (K_{11})^{1/2} \\ L_{ii} = \left(K_{ii} - \sum_{k=1}^{i-1} L_{ik}^2\right)^{1/2} \qquad i = 2, \ldots, n-1 \\ L_{ij} = \dfrac{K_{ij} - \sum_{k=1}^{j-1} L_{ik}L_{kj}}{L_{jj}}, \qquad \forall\ i \rangle j \\ L_{nn} = (K_{nn})^{1/2} \end{cases} \qquad (3.17)$$

The algorithm for Cholesky decomposition is presented in Table 3.4.

Table 3.4 Algorithm for Cholesky decomposition

Loop 1 i = 1 until n − 1
    $K_{ii} = K_{ii}^{1/2}$
    Loop 2 j = k+1 until n
        $K_{ji} = K_{ji} / K_{ii}$
    Endloop 2
    Loop 3 k = i+1 until n
        Loop 4 m = k until n
            $K_{mk} = K_{mk} - K_{mi}K_{ki}$
        Endloop 4
    Endloop 3
Endloop 1
$K_{nn} = K_{nn}^{1/2}$

60 Solution of Simultaneous Linear Equations

Note that to save storage the lower triangular matrix is stored in the locations of the matrix **K** below its diagonal. The elements of $\mathbf{L}^T$ are readily obtained directly from **L**:

$$\mathbf{L}^T_{ij} = \mathbf{L}_{ji} \tag{3.18}$$

Similar to LU decomposition, Cholesky decomposition consists of three steps. The first step is the decomposition of **K**. The second step is the forward solution of Eq. (3.7) to find **v**. The final step is the backward solution of Eq. (3.6) for **u** where $\mathbf{U} = \mathbf{L}^T$.

3.4 INDIRECT METHODS

The indirect methods for solution of simultaneous linear equations have their roots in the numerical solution of unconstrained nonlinear optimization problems. One of the simplest indirect methods is *the steepest descent* method. In this approach, the problem of solution of the simultaneous linear equations represented by Eq. (3.1) is transformed to the minimization of the following function:

$$F(\mathbf{x}) = \frac{1}{2}\mathbf{x}^T \mathbf{K} \mathbf{x} - \mathbf{x}^T \mathbf{F} \tag{3.19}$$

where $\mathbf{K} \in R^{n \times n}$ is a symmetric positive definite matrix, $\mathbf{F} \in R^n$ and R^n is an n-dimensional space.

The gradient of the function F is

$$\nabla F(\mathbf{x}) = \mathbf{K}\mathbf{x} - \mathbf{F} \tag{3.20}$$

When the function F is minimum, this gradient is zero and Eq. (3.20) yields **x** = **u** which is the exact solution of the simultaneous linear equations represented by Eq. (3.1).

In indirect or iterative methods, the minimum solution of Eq. (3.19) is obtained iteratively. In each iteration k, the values of the vector **x** are updated by

$$\mathbf{x}_{k+1} = \mathbf{x}_k + \Delta\mathbf{x}_k \tag{3.21}$$

where $\Delta\mathbf{x}_k$ represents a small change to the vector $\mathbf{x}_k$ with the objective of reducing the function F. In the optimization parlor, function F is called the objective function.

In the steepest descent method, the incremental change vector $\Delta\mathbf{x}_k$ at the kth iteration is taken as the product of a direction search vector $\mathbf{r}_k$ and a step length α_k:

$$\mathbf{x}_{k+1} = \mathbf{x}_k - \alpha_k \mathbf{r}_k \tag{3.22}$$

The direction search vector is taken as the negative of the gradient of the objective function

$$\mathbf{r}_k = -\nabla F(\mathbf{x}_k) \tag{3.23}$$

Substituting Eq. (3.22) into Eq. (3.19) yields

$$F(\mathbf{x}_{k+1}) = \frac{1}{2}(\mathbf{x}_k - \alpha_k \mathbf{r}_k)^T \mathbf{K}(\mathbf{x}_k - \alpha_k \mathbf{r}_k) - (\mathbf{x}_k - \alpha_k \mathbf{r}_k)^T \mathbf{F} \tag{3.24}$$

Now, we minimize the objective function F with respect to the step length α_k as a one-dimensional search.

$$\frac{dF(\mathbf{x}_{k+1})}{d\alpha_k} = \mathbf{r}_k^T \mathbf{K} \mathbf{r}_k \alpha_k - \mathbf{r}_k^T (\mathbf{K}\mathbf{x}_k - \mathbf{F}) = 0 \tag{3.25}$$

From Eq. (3.25), we find

$$\alpha_k = \frac{\mathbf{r}_k^T \mathbf{r}_k}{\mathbf{r}_k^T \mathbf{K} \mathbf{r}_k} \tag{3.26}$$

The steps involved in the steepest descent method are presented in Table 3.5.

Table 3.5: Steepest Descent Algorithm

Step 1: For k =0, select initial vector $\mathbf{x}_0 = 0$, find residual vector $\mathbf{r}_0 = \mathbf{F} - \mathbf{K}\mathbf{x}_0$, and select the convergence tolerance ε

Step 2: Compute α_k to minimize $F(\mathbf{x}_k - \alpha_k \mathbf{r}_k)$

$$\alpha_k = \frac{\mathbf{r}_k^T \mathbf{r}_k}{\mathbf{r}_k^T \mathbf{K} \mathbf{r}_k}$$

Step 3: Update the solution vector $\mathbf{x}_{k+1} = \mathbf{x}_k - \alpha_k \mathbf{r}_k$

Step 4: Update the gradient vector $\mathbf{r}_k = -\mathbf{K}\mathbf{x}_{k+1} + \mathbf{F}$

Step 5: Check the convergence, if $\|\mathbf{r}_{k+1}\|_2^2 \langle \varepsilon$, stop; otherwise continue

Step 6: Set k = k+1 and go to step 2

3.5 CONJUGATE GRADIENT DIRECTION METHOD

The steepest descent method is a simple indirect method for solution of the simultaneous linear equations. But, its convergence rate is slow when the ratio of the maximum to minimum eigenvalues of matrix **K** is large (Golub and Van Loan, 1991). Various methods have been proposed to improve the convergence of this approach, mostly through judicious selection of the search direction. The simplest modification of the steepest descent method was proposed by Fletcher and Reeves (1964). Suppose the search direction vector for the kth iteration is $\mathbf{p}_k$. The initial direction search vector is the same as the steepest descent method. However, for the subsequent iterations, the direction search vector is modified as follows:

$$\mathbf{p}_{k+1} = \mathbf{r}_{k+1} + \beta_{k+1}\mathbf{p}_k \tag{3.27}$$

where

$$\beta_{k+1} = \frac{\mathbf{r}_{k+1}^T \mathbf{r}_{k+1}}{\mathbf{r}_k^T \mathbf{r}_k} \tag{3.28}$$

The steps involved in the conjugate gradient direction method are given in Table 3.6. In the conjugate gradient direction method the exact solution is obtained with fewer than n iterations (Ortega, 1988), where n is the number of unknowns (the number of nodal degrees of freedom in the structural analysis problem). When a good approximation for initial solution is available, this method usually converges quickly to an accurate solution with relatively a few iterations. However, for some problems, the conjugate gradient direction method shows very slow convergence. In order to improve the rate of convergence, the Preconditioned Conjugate Gradient method has been introduced.

Table 3.6: Conjugate Gradient Direction Method

Step 1: Initialize the solution vector $\mathbf{x}_0 = 0$, the gradient vector $\mathbf{r}_0 = \mathbf{F} - \mathbf{K}\mathbf{x}_0$, the conjugate direction vector $\mathbf{p}_0 = \mathbf{r}_0$, select the convergence tolerance ε and set k = 1.

Step 2: Compute α_k to minimize $F(\mathbf{x}_k - \alpha_k \mathbf{p}_k)$

$$\alpha_k = \frac{\mathbf{r}_k^T \mathbf{r}_k}{\mathbf{p}_k^T \mathbf{K} \mathbf{p}_k}$$

Step 3: Update the solution vector $\mathbf{x}_{k+1} = \mathbf{x}_k - \alpha_k \mathbf{p}_k$.

Step 4: Update the gradient vector $\mathbf{r}_{k+1} = \mathbf{r}_k - \alpha_k \mathbf{K} \mathbf{p}_k$.

Step 5: Check the convergence, if $\|\mathbf{r}_{k+1}\|_2^2 \leq \varepsilon$, stop

Step 6: Compute the coefficient β_{k+1} for updating the conjugate direction vector

$$\beta_{k+1} = \frac{\mathbf{r}_{k+1}^T \mathbf{r}_{k+1}}{\mathbf{r}_k^T \mathbf{r}_k}$$

Step 7: Compute the new conjugate direction vector
$$\mathbf{p}_{k+1} = \mathbf{r}_k + \beta_{k+1}\mathbf{p}_k$$

Step 8: Set k = k + 1 and go to step 2.

3.6 PRECONDITIONED CONJUGATE GRADIENT METHOD

The main idea behind preconditioning is to find a matrix $\mathbf{K}'$ that can reduce the ratio between the largest to the smallest eigenvalues of matrix $\mathbf{K}$. As mentioned earlier, a large value for this ratio will result in a slow rate of convergence. From matrix $\mathbf{K}'$, the new direction vector $\bar{\mathbf{r}}$ is obtained by solving $\mathbf{K}'\bar{\mathbf{r}} = \mathbf{r}$. At each iteration the new gradient vector from the preconditioned matrix must be solved. Therefore, the selection of the preconditioning matrix approximating matrix $\mathbf{K}$ causes a dilemma. If the matrix $\mathbf{K}'$ is close to matrix $\mathbf{K}$, the processing time to compute $\bar{\mathbf{r}}$ from $\mathbf{K}'\bar{\mathbf{r}} = \mathbf{r}$ is almost the same as that to solve $\mathbf{Ku} = \mathbf{F}$. The simplest preconditioning matrix is a diagonal matrix, which is also known as Jacobi preconditioning.

The PCG method converges to the exact solution faster than the conjugate gradient direction method. In terms of memory storage, the PCG method requires 6n additional memory storage. However, this additional memory storage is relatively small compared with the memory needed to store matrix $\mathbf{K}$. Various types of preconditioning and their convergence properties are covered in many references such as

Golub and Van Loan, (1991), Ortega, (1989), and Luenberger, (1984).

The PCG method can compete with direct methods such as the LU and Cholesky decompositions when the matrix has a large bandwith as shown in Chapter 5. The steps of the PCG method based on diagonal preconditioning are presented in Table 3.7.

Table 3.7: The Preconditioned Conjugate Gradient Method

Step 1: Initialize the solution vector $\mathbf{x}_o = 0$, and the gradient vector $\mathbf{r}_o = \mathbf{F} - \mathbf{K}\mathbf{x}_o$.

Step 2: Determine the preconditioning matrix $\mathbf{K}'$ which is symmetric and positive definite.

Step 3: Compute the new gradient vector $\bar{\mathbf{r}}_o$ from $\mathbf{K}'\bar{\mathbf{r}}_o = \mathbf{r}_o$, set $\mathbf{p}_o = \bar{\mathbf{r}}_o$, select the convergence tolerance ε, and set k = 1.

Step 4: Compute α_k to minimize $F(\mathbf{x}_k - \alpha_k \mathbf{p}_k)$

$$\alpha_k = \frac{\bar{\mathbf{r}}_k^T \mathbf{r}_k}{\mathbf{p}_k^T \mathbf{K} \mathbf{p}_k}$$

Step 5: Update the solution vector $\mathbf{x}_{k+1} = \mathbf{x}_k - \alpha_k \mathbf{p}_k$.

Step 6: Update the gradient vector $\mathbf{r}_{k+1} = \mathbf{r}_k + \alpha_k \mathbf{K} \mathbf{p}_k$.

Step 7: Check the convergence, if $\|\mathbf{r}_{k+1}\|_2^2 \leq \varepsilon$, stop; otherwise continue.

Step 8: Compute the new gradient vector from $\mathbf{K}'\bar{\mathbf{r}}_{k+1} = \mathbf{r}_{k+1}$.

Step 9: Compute the coefficient β_k for updating the conjugate direction vector.

$$\beta_{k+1} = \frac{\bar{\mathbf{r}}_{k+1}^T \mathbf{r}_{k+1}}{\bar{\mathbf{r}}_k^T \mathbf{r}_k}$$

Step 10: Compute the new conjugate direction vector.

$$\mathbf{p}_{k+1} = \bar{\mathbf{r}}_k + \beta_{k+1}\mathbf{p}_k$$

Step 11: Set k = k + 1 and go to step 4.

3.7 GLOBAL STIFFNESS MATRIX

In the structural analysis problem, the matrix **K** in Eq. (3.1) is obtained by assembling the member stiffness matrices into the global stiffness matrix. The solutions of systems of linear equations presented in the previous sections are based on a full **K** matrix. However, the nonzero entries of the global stiffness matrices often are clustered near the diagonal in a banded form. The number of computations can be reduced substantially by taking into account the banded form of the global stiffness matrix, and operating only on the entries within the banded region.

In order to illustrate the process of assembling the member stiffness matrices, let us consider the three-bar truss with eight degrees of freedom as shown in Figure 3.1. The node numbering system in Figures 3.1a and 3.1b is interchanged for nodes 1 and 2. The number of nodes in this example is 4 and each node has two degrees of freedom. This makes the size of the global stiffness matrix 8x8. The largest difference of the node numbers for a member based on the nodes arrangement of Figure 3.1a is 3 which is for member 3.

Each member has a 4x4 stiffnesses matrix and the entry numbers of the member stiffnesses based on the corresponding degrees of freedom are shown in Figure 3.2. These member stiffness matrices are assembled into the global stiffness matrix shown in Figure 3.3a. The global stiffness matrix is symmetric and positive definite. In order to reduce the storage and computer processing time, only the upper triangular of the global stiffness matrix is stored in a banded form as shown in

a) b)

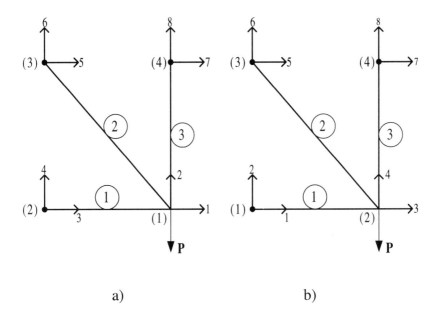

Figure 3.1 : A three-bar truss

11	12	13	14
21	22	23	24
31	32	33	34
41	42	43	44

Member stiffness
member 1

11	12	15	16
21	22	25	26
51	52	55	56
61	62	65	66

Member stiffness
member 2

11	12	17	18
21	22	27	28
71	72	77	78
81	82	87	88

Member stiffness
member 3

Figure 3.2 : Entry numbers of member stiffness matrices based on the nodes arrangement in Figure 3.1a

Figure 3.3b. The diagonal entries in the full matrix become the first column entries in the banded form. The bandwidth of the new storage arrangement is 8. Therefore, there is no reduction in memory storage.

The dimension of the global stiffness matrix **K** can be reduced by rearranging the node numbering scheme. Let us consider the node numbering example in Figure 3.1b. The largest difference of the node numbers is 2 (for member 3). Each member has a 4x4 stiffness matrix and the entry numbers of the member stiffness matrix based on the corresponding degrees of freedom are shown in Figure 3.4. These member stiffness matrices are assembled into the global stiffness matrix shown in Figure 3.5a. The global stiffness matrix stored in a banded form is shown in Figure 3.5b. For the node numbering scheme shown in Figure 3.1b, the bandwidth, mb, is 6 and the size of the banded stiffness matrix is 8x6, resulting in a smaller storage requirement compared with the node numbering scheme shown in Figure 3.1a. For large problems, the reduction in storage as well as the computer processing time becomes very significant.

The bandwidth can be computed from the following formula:

$$\text{mb} = \max(\text{ABS}(\text{node i} - \text{node j}) + 1) * \text{ndfn}$$

where nodes i and j are the two node numbers of a member, and ndfn is the number of degrees of freedom per node.

In a large structure such as high-rise buildings minimizing the bandwidth will result in minimizing the required storage and processing time. The banded system Cholesky method and banded storage Cholesky method when the matrix **K** is stored in banded form are presented in Tables 3.8 and 3.9, respectively. In these tables, ib = mb − 1.

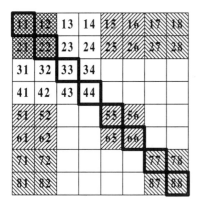
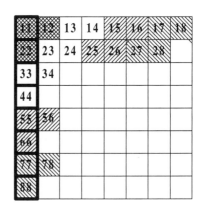

a) Full matrix

b) Upper triangular matrix stored in banded form

☐ entry for member 1   ▨ entry for member 2   ▧ entry for member 3

Figure 3.3: Global stiffness matrix based on the node arrangement in Figure 3.1a

11	12	13	14
21	22	23	24
31	32	33	34
41	42	43	44

Member stiffness member 1

33	34	35	36
43	44	45	46
53	54	55	56
63	64	65	66

Member stiffness member 2

33	34	37	38
43	44	47	48
73	74	77	78
83	84	87	88

Member stiffness member 3

Figure 3.4 : Entry numbers for member stiffness matrices based on the node arrangement in Figure 3.1b.

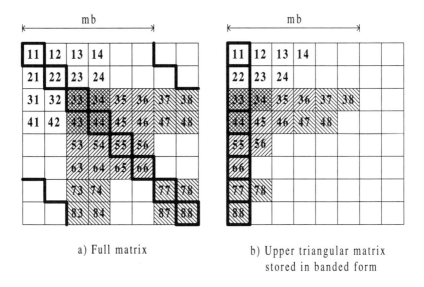

a) Full matrix   b) Upper triangular matrix stored in banded form

☐ entry for member 1   ▨ entry for member 2   ▧ entry for member 3

Figure 3.5: Global stiffness matrix based on the node arrangement in Figure 3.1b.

Table 3.8: Algorithm for banded Cholesky decomposition

Loop 1 $i = 1$ until $n-1$
    $K_{ii} = K_{ii}^{1/2}$
    Loop 2 $j = i+1$ until $\min(i+ib, n)$
        $K_{ji} = K_{ji} / K_{ii}$
    Endloop 2
    Loop 3 $k = i+1$ until $\min(i+ib, n)$
        Loop 4 $m = k$ until $\min(i+ib, n)$
            $K_{mk} = K_{mk} - K_{mi} K_{ik}$
        Endloop 4
    Endloop 3
Endloop 1
$K_{nn} = K_{nn}^{1/2}$

Table 3.9 Algorithm for Cholesky decomposition with banded storage

Loop 1 i = 1 until n − 1
    $K_{i1} = K_{i1}^{1/2}$
    Loop 2 j = 2 until min (mb, n − i + 1)
        $K_{ij} = K_{ij} / K_{i1}$
    Endloop 2
    im = ib+1
    jm = 1
    Loop 3 k = i+1 until min (i+ib, n)
        im = im − 1
        jm = jm + 1
        t = jm − 1
        Loop 4 m = 1 until min (im, n − 1)
            km = t + m
            $K_{km} = K_{km} - K_{i,km} K_{i,jm}$
        Endloop 4
    Endloop 3
    jm = 1
Endloop 1
$K_{n1} = K_{n1}^{1/2}$

CHAPTER 4
Vectorization Techniques

4.1 VECTOR PROCESSING

A review of high-performance computing methods in structural mechanics is presented by Adeli et al. (1993). Parallel machines that have vector processors are called parallel-vector computers. The Cray YMP8/864 is one such machine. This computer has eight central processing units (CPU). Each CPU has random access to the main shared memory.

There are two primary approaches to improve the performance and efficiency of computations on a parallel-vector machine: vector processing and parallel processing. It will be shown that on machines with a few processors such as the Cray YMP8/864 with eight processors, the speedup contribution of vector processing can be substantially more than parallel processing. The architecture of the register block diagram of the

Cray YMP8/864 is shown in Figure 4.1 (Cray, 1991). Broadly speaking, each processor consists of four parts: shared memory, sets of registers, instruction buffer, and sets of functional units. Each processor has five sets of registers: vector, scalar, address, T-, and B-registers. Vector, scalar, and address registers have direct access to the functional unit and the primary operating registers are vector and scalar registers. The T- and B-registers are used as backup registers when the corresponding address and scalar registers are full. Address registers are used as memory references, to count the loops, to provide shift counts, and as an index register.

The shared memory is divided into 256 banks and each processor has random access to the shared memory as shown in Figure 4.2 (Cray, 1991). In addition, each processor has its own private memory. The CPU cycle-time is equal to six nanoseconds and memory access time is equal to five CPU cycle-times or 30 nanoseconds. The main memory with 256 banks can transfer $256/30 \times 10^{-9} = 8.53 \times 10^9$ words per second. The ability of one processor to fetch one word is called the fundamental period. In the Cray YMP8/864, it is equal to six nanoseconds $(6 \times 10^{-9} \text{ seconds})$ or one processor can fetch 0.1667×10^9 words per second. If one processor can fetch two words simultaneously, the total number of words per second becomes 0.333×10^9. Theoretically, one processor can perform 333×10^6 operations per second. However, in reality, it is much lower than 333×10^6 operations due to several reasons such as the non-vectorizable portion of the code, overhead, etc. Increasing the vectorizable parts of the code is an important part of developing efficient algorithms and codes.

Vector processing on vector computers is performed by executing different parts of an arithmetic instruction simultaneously in the form of vectors using the idea of

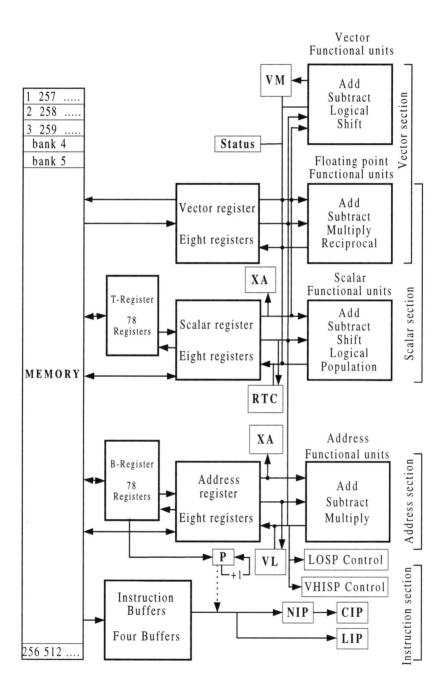

Figure 4.1: The Architecture of Cray YMP8/864

76 Vectorization Techniques

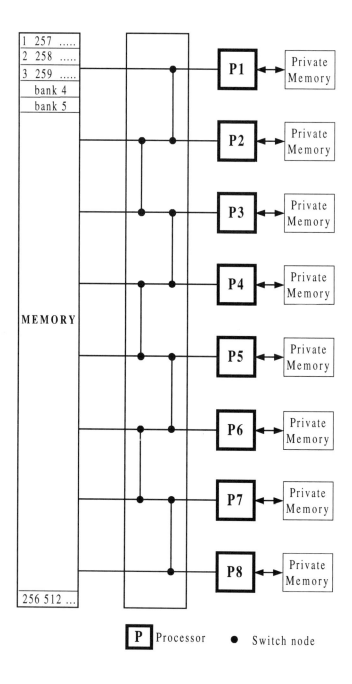

Figure 4.2: Random access to the shared and private memories

pipelining. The concept of pipelining is similar to the process of assembling a car where different parts of the car are assembled at the same time.

In arithmetic terms, suppose there are M operations and each operation is divided into N steps. The execution of M arithmetic operations in a pipelining fashion is shown in Figure 4.3. The time lags in performing M different operations shown in Figure 4.3 are due to fetching data and other overhead factors. But, this time is only a small fraction of the time needed to actually perform the operation. Let the time needed to execute a single arithmetic operation be t_1 and the time used to complete M arithmetic operations in a pipelining fashion be t_M. The average time used to execute one arithmetic operation in a vector fashion is t_M / M which is a fraction of t_1. Because the time lag between subsequent operations is very small, M different operations are executed virtually simultaneously.

Consider the vector arithmetic operation of adding vectors **A** and **B** and storing the results in vector **C** (Figure 4.4). On the Cray YMP8/864, this operation can be performed in two different ways. Either as a memory-to-memory process where the data are transferred from the main memory to the functional units performing the operations and the results are transferred back to the main memory. Or, as a register-to-register process where the data are transferred from the vector register to the functional units and the results are stored in the vector registers. The register-to-register process is faster than the memory-to-memory process. Consequently, every attempt has to be made to perform the operations as register-to-register in order to maximize the vectorization speedup of the code.

The Cray YMP8/864 machine provides eight vector registers with each vector register capable of storing 64

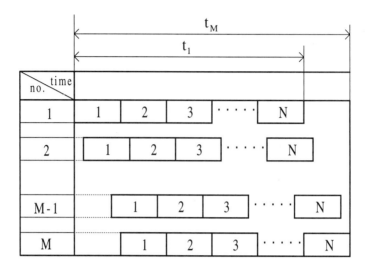

Figure 4.3: The execution of M operations in a pipelining fashion

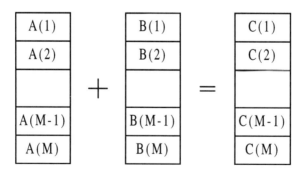

Figure 4.4: Vector pipelining for the addition operation

elements (V1 to V8 in Figure 4.5). As such, up to 64 elements can be processed almost simultaneously. If the number of elements in a vector such as **A** or **B** in the aforementioned addition operation is more than 64, the first 64 elements are operated on first and stored in a vector register, to be followed by the next 64 elements, and so on.

The vector processing of the arithmetic operation **AxB+C=D** is shown in Figure 4.6. Vectors **A** and **B** are streamed from the vector registers V1 and V2 into the multiplication functional unit to produce **AxB**. The results are directly added to **C** streaming from the vector register V3 to produce the final outcome stored in vector register V4. This operation is called vector chaining.

4.2 VECTORIZATION TECHNIQUES

There are basically two approaches to enhance the vectorization performance. The first approach is simply to utilize the capability of compiler technology. Using this approach alone, however, significant improvement on the performance of computation cannot be achieved. The second approach is to modify or extend the portions of the code that are considered non-vectorizable to satisfy the vectorization requirement. The Cray YMP 8/864 compiler vectorizes the inner DO loop automatically (Cray, 1990). Therefore, we have to make the inner DO loops vectorizeable as much as possible while keeping the number of synchronization points at a minimum. In the second approach, one must attempt to remove the barriers that inhibit vectorization in the DO loop in order to enhance the performance of vectorization. Statements in the DO loop may depend on another statement from the same iteration or from another iteration. This is called the dependency

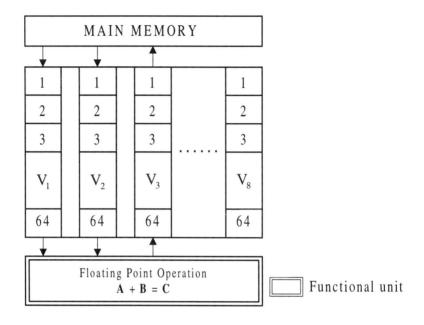

Figure 4.5: Register to register addition

relationship. Four different types of dependency can be recognized as shown in the example of Figure 4.7. The first type is control dependency (CD) where the execution of statements S_2 or S_3 depends on the true or false status of statement S_1. Control dependency determines the order of execution of statements.

The second type of dependency is flow dependency (FD). For example, the execution of statement S_3 depends on the value assigned to statement S_2. The third type of dependency is anti-dependency (AD). For example, the value of D(i) is assigned in statements S_4 and S_5 and used in the statements S_2 and S_3. The relationship between S_4 and S_5 is called output dependency (OD). The value of D(i+1) is assigned in statement

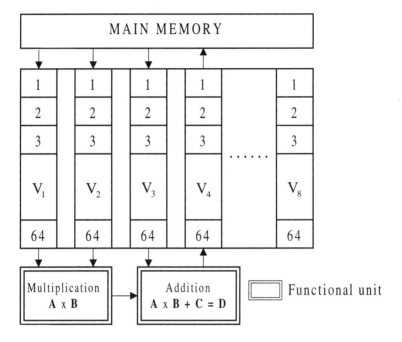

Figure 4.6: Vector chaining

S_4 as well as in statement S_5. In output dependency, interchanging the order of S_4 and S_5 results in error.

To maximize the vectorization in a code, we need to establish the so-called strongly connected components in a flow graph (Zima and Chapman, 1991). A strongly connected component is the portion of the flow graph with a closed cycle. Each strongly connected component is programmed as a DO loop in the code and can be vectorized. After the strongly connected components have been identified, we should search for dependencies and circumvent them as much as possible in order to maximize vectorization.

We identify and describe seven methods for enhancing

Loop 1 i = 1 until n
S1   : If (A(i). GT.0) go to 10
S2   : B(i) = D(i) + 2
10 S3   : C(i) = B(i) + D(i)
S4   : D(i+1) = E(i)
....
....
S5   : D(i+1) = B(i) * 2
Endloop 1

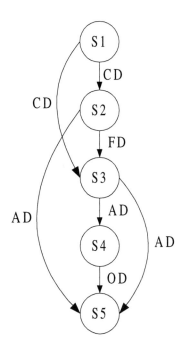

Figure 4.7: Dependency graph

the vectorization. The first method is to remove the inhibitors that break the flow control, such as input/output statements, subroutine or function calls and return or stop statements. The second method is subroutine inlining, that is, a subroutine is transferred into the DO loop. This allows the DO loop to be vectorized and the overhead for transferring control to a subroutine is avoided.

The third method is to eliminate dependencies. This is one of the most effective ways of improving the performance of vectorization. There are several methods to eliminate the dependency, depending on the problem type. For instance, a DO loop can be divided into two loops, one vectorizable and the other non-vectorizable. This is called loop splitting. Another way is to interchange the statements. This is called loop reordering.

The fourth method is to eliminate short inner DO loops by repeating the statements comprising the loop (loop unrolling vertically) or adding the statements that comprise the loop (loop unrolling horizontally) so that the outer DO loop or the surrounding loop can be vectorized.

The fifth method is to extend the dimension of the variables. Sometimes loop splitting cannot be performed because the same variables used in one statement are also used in another statement in a DO loop. In such a case, the results from the first statement are stored in extended variables.

The sixth method is to reduce the time for memory access. The data stream from the main memory to the functional unit (where the arithmetic operations are performed) plays an important role in vectorization. The data stream from the same bank within the bank-cycle time will deteriorate the performance of vectorization. Therefore, the data stream at the inner DO loop has to be arranged columnwise for the FORTRAN compiler.

The final method is to change or modify the algorithm. When all the necessary steps toward the enhancement of vectorization have been taken, and the performance of vectorization is still not satisfactory, we may have to resort to this approach.

CHAPTER 5
Parallel-Vector Algorithms for Analysis of Large Structures

5.1 INTRODUCTION

The computational cost of the structural analysis is often dominated by the cost of assembling element stiffness matrices and solving the resulting system of linear equations. Current research has been directed towards the development of algorithms to solve systems of simultaneous linear equations by optimizing the capabilities of vector, parallel or parallel-vector computers (Poole and Overman, 1988; Ortega, 1989; Agarwal, et al., 1990; Storaasli, et al., 1990; Hsu and Adeli, 1991; Shivakumar, et al., 1992; Adeli and Kamal (1992 a&b)). A survey of parallel machines is presented by Adeli and Visnubholta (1992).

In this chapter, parallelization and vectorization of assembling the stiffness matrix in the structural analysis problem and solution of systems of simultaneous linear equations are studied. Both direct and indirect methods for solution of systems of simultaneous linear equations are investigated. The performance of parallel-vector algorithms for solving large structural analysis problems on Cray YMP8/864 is presented.

5.2 CONCURRENT PROCESSING

Concurrent processing on a MIMD machine like the Cray YMP8/864 is performed by multitasking, that is dividing the program into a number of independent tasks and executing them on various processors simultaneously. Multitasking is divided into macrotasking, microtasking, and autotasking. Macrotasking is multitasking at the subroutine level. Each processor executes an entire subroutine independently.

Microtasking is multitasking at the DO loop level. While vectorization is done in the inner DO loop, microtasking is performed in the outer DO loop by the CF77 precompiler. Autotasking is the automatic distribution of tasks to processors by the compiler. Autotasking works best in programs where most of the code consists of nested loops. Autotasking cannot be performed when loop iterations contain interdependent array elements or when there is a function call (Saleh and Adeli, 1994).

In a multitasked program, two types of tasks are identified, the first is a master task (executed on the master processor) which executes all of the sequential codes and the multitasked loops. The second is slave tasks (executed on slave processors) which only execute the multitasked loop. When a parallel region is entered, the master processor calls the slave processors to execute the multitasked loop. The private memory in the master processor is copied to the private memory of each slave processor. The master processor also performs the initialization and termination of the parallel region. Hence, the master processor has to wait until all the slave processors complete the multitasked loop before continuing to the next parallel region or sequential part of the algorithm. The execution of a multitasked code on a shared-memory machine is shown in Figure 5.1.

5.3 CONCURRENT EVALUATION AND ASSEMBLY OF THE STRUCTURE STIFFNESS MATRIX

In a structural analysis problem using the displacement (stiffness) method, the major computational steps are evaluation and assembly of the element stiffness matrices and solution of the resulting system of simultaneous linear equations for nodal displacements. Therefore, in this chapter we focus our attention on these two steps. We limit the scope of our presentation to space axial-load structures. The element stiffness matrix for these structures is a 6×6 square matrix. While there is inherent parallelism in the evaluation of element stiffness matrices, the assembly of element stiffness matrices into the global structure stiffness matrix (**K**) is inherently sequential. This situation arises because of the contention for the data corresponding to the nodes shared by more than one element. To avoid undesirable racing conditions and to allow vectorization, we enlarge the dimensions of the element stiffness matrix from **k**(6,6) to **k**(6,6,M) where M is the number of elements in the structure and we employ the directives *Guard* and *Endguard* provided by the Cray YMP8/864 FORTRAN compiler.

88 Parallel-Vector Algorithms for Analysis of Large Structures

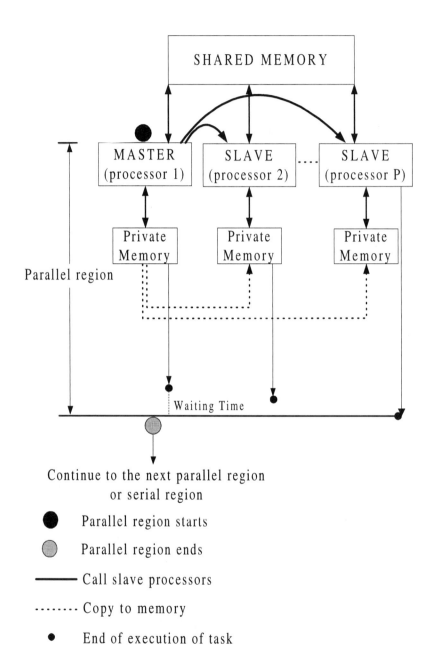

Figure 5.1: Execution of multitasked code

5.4 CONCURRENT CHOLESKY AND LU DECOMPOSITION METHODS

Solution of systems of linear simultaneous equations to solve the nodal displacements is the most time-consuming part in structural analysis, especially for large structures. In this section, we focus on the LU and Cholesky decomposition methods for solving a system of linear simultaneous equations. For the sake of brevity, only the parallel-vector algorithm for solution of a system of linear simultaneous equations using the Cholesky decomposition is presented. The system of linear simultaneous equations to be solved is represented by Eq. (3.1). As noted in Section 3.3, the Cholesky method consists of two parts. The first part is factorization of matrix **K** into the product of lower and upper triangular matrices **L** and $\mathbf{L}^T$. The second part is the solution of the vector **u** through forward and backward substitutions.

The parallel-vector algorithm for Cholesky factorization is given in Table 5.1. Most of the computational time is spent in the factorization of matrix **K**. Basically there are two approaches to enhance the performance of Cholesky factorization on a parallel-vector machine. The first is to maximize the capability of vector compiler technology. Significant improvement can be obtained through minimizing the CPU time for data stream from the main memory to the functional unit (where the arithmetic operations are performed). Further improvement can be achieved through maximizing the capability of compiler to perform arithmetic operations in the inner loop by employing loop unrolling.

Based on the ordering of the loops, the Cholesky factorization can be performed in six different ways. In order to minimize the required CPU time, the *kji* form as illustrated in

Table 5.1 Parallel-vector Cholesky factorization for an n x n matrix with semi-bandwidth of b

Loop 1 k = 1 until n−1
   $K_{kk} = K_{kk}^{1/2}$
   Loop 2 m = k+1 until min (k+b,n)
      $K_{mk} = K_{mk} / K_{kk}$
   Endloop 2
   Loop 3 j = k+1 until min (k+b,n) (*microtasking*)
      Loop 4 i = j until min (k+b,n) (*vectorization*)
         $K_{ij} = K_{ij} - K_{ik} K_{jk}$
      Endloop 4
   Endloop 3
Endloop 1
$K_{nn} = K_{nn}^{1/2}$

Table 5.1 is chosen for a shared memory machine such as the Cray YMP8/864 and implemented in FORTRAN. In this algorithm the CPU-intensive operations are inside the inner loop 4. The elements K_{ik} and K_{jk} stream from the main memory to each functional unit (addition, subtraction, multiplication, and reciprocal), with a constant stride of one (K_{1k}, K_{2k}, K_{3k}, K_{4k},...). Stride of a vector is an increment of the index between two consecutive data stored in the main memory columnwise. The results from the multiplication operation $K_{ik} * K_{jk}$ are used directly in another operation (subtraction). Thus, the CPU time is reduced and the memory bank conflict is avoided entirely.

Further improvement is achieved by adding operations to loop 4 (Table 5.1) through loop unrolling horizontally (Poole and Overman, 1988). Our improved parallel-vector algorithm for Cholesky factorization is presented in Table 5.2. Before

Table 5.2 Improved parallel-vector algorithm for the banded Cholesky factorization

Loop 1 k = 1 until n − s − 1,s
   Loop 2 o = k until k+s-2
      $K_{oo} = K_{oo}^{1/2}$
      Loop 3 s = k+1 until min (k+b,n)
         $K_{so} = K_{so} / K_{oo}$
      Endloop 3
      Loop 4 j = o+1 until k+s-1
         Loop 5 i = j until min (o+b,n) (*vectorization*)
            $K_{ij} = K_{ij} - K_{io} K_{jo}$
         Endloop 5
      Endloop 4
   Endloop 2
   c = k+s − 1
   $K_{cc} = K_{cc}^{1/2}$
   Loop 6 t = c+1 until min (c+b,n)
      $K_{tc} = K_{tc} / K_{cc}$
   Endloop 6
   u = k
   Loop 7 v = k+b+1 until min (k+b+s − 1,n)
      u = u +1
      in = min (k+s − 1,n)
      im = u − in − s1 − 1
      if (im.ge.s1) then
         q = (im/s1 − 1)*s1
         Loop 8 w = u until u+1,s1
            Loop 9 x = k+s until v (*vectorization*)
               $K_{vx} = K_{vx} - K_{vw} K_{xw} ..... - K_{vw+s-1} K_{xw+s-1}$
            Endloop 9
         Endloop 8

Table 5.2-continued

    Loop 10 y = u+q+1 until in
      Loop 11 z = k+s until v (*vectorization*)
$$K_{vz} = K_{vz} - K_{vy}K_{zy}$$
      Endloop 11
    Endloop 10
  else
    Loop 12 w = u until in
      Loop 13 x = k+s until v (*vectorization*)
$$K_{vx} = K_{vx} - K_{vw}K_{xw}$$
      Endloop 13
    Endloop 12
  Endif
Endloop 7
Loop 14 u = 1 until P (*microtasking*)
$u_1 = (u-1)*(b-s+1)/2P+1$
$u_2 = u*(b-s+1)/2P$
  Loop 15 j = u_1 until u_2
    Loop 16 i = j until min(k+b,n) (*vectorization*)
$$K_{ij} = K_{ij} - K_{ik}K_{jk} - .... - K_{ik+s-1}K_{jk+s-1}$$
    Endloop 16
  Endloop 15
$u_3 = (p-u)*(b-s+1)/2P+1$
$u_4 = (p-u+1)*(b-s+1)/2P$
  Loop 17 m = u_3 until u_4
    Loop 18 d = m until min(k+b,n) (*vectorization*)
$$K_{dm} = K_{dm} - K_{dk}K_{mk} - ..... - K_{dk+s-1}K_{mk+s-1}$$
    Endloop 18
  Endloop 17
Endloop 14

Table 5.2-continued

Endloop 1
c=c+1
   Loop 19 k = 1 until n−1
      $K_{kk} = K_{kk}^{1/2}$
      Loop 20 l = k+1 until min (k+b,n)
         $K_{lk} = K_{lk} / K_{kk}$
      Endloop 20
      Loop 21 j = k+1 until n (*microtasking*)
         Loop 22 i = j until n (*vectorization*)
            $K_{ij} = K_{ij} - K_{ik} * K_{jk}$
         Endloop 22
      Endloop 21
   Endloop 19
   $K_{nn} = K_{nn}^{1/2}$

horizontal loop unrolling is performed in loops 12 and 14, columns and rows that are involved in the unrolling process are updated in loops 2 through 6. The procedure for updating these columns and rows is shown schematically in Figure 5.2. For instance, in order to perform loop unrolling in the (s+1)st loop, columns s+1 to 2s and rows b+s+2 to b+2s have to be updated where s is the loop increment (the depth of loop unrolling). Loops 5, 9, 11, 16, 18, and 22 in Table 5.2 are equivalent to loop 4 in Table 5.1. Another loop unrolling is performed in loop 9, where the rows are updated. In the second loop unrolling, the loop increment, s1, is smaller than s because the size of the update rows is s. When s is large, the size of the rest of the matrix to be updated, $b-s$, becomes small. Consequently, less

94 Parallel-Vector Algorithms for Analysis of Large Structures

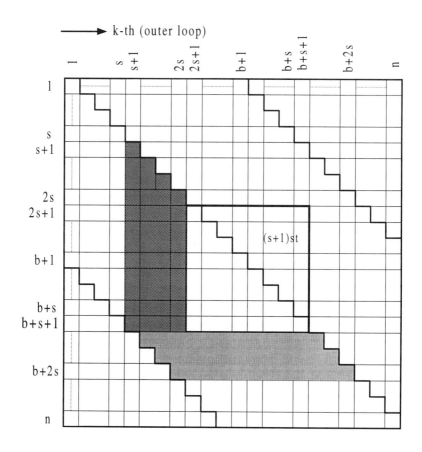

▩ Column update at the (s+1)st outer Loop
▢ Row update at the (s+1)st outer Loop
s = outer loop increment
b = semi-bandwidth
n = number of equations

Figure 5.2: Column and row updating in improved parallel-vector algorithm for the Cholesky factorization

benefit is obtained from loop unrolling. On the other hand, when s is small the vector chaining capability is not optimized. In this work, we found s = 9 and s1 = 2 result in good performance for Cholesky factorization.

Another approach to enhance the performance of Cholesky factorization is to exploit parallelism and employ multitasking. The computation in loop 3 of Table 5.1 should be distributed among the processors as equally as possible. The algorithm presented in Table 5.1 suffers from load imbalance, because at the kth outer loop in loop 3, each processor is assigned the same number of columns with each column representing a task. But these tasks have variable lengths from the longest j = b to the shortest at j = 1 (Figure 5.3). To improve the algorithm given in Table 5.1, the tasks in loop 3 are divided into 2P tasks, P being the number of processors. Each processor is assigned a combination of long and short tasks, as indicated in Figure 5.3c. Loops 4, 15, 17, and 21 of Table 5.2 are equivalent to loop 3 of Table 5.1. The balancing of tasks is performed in loops 14 through 18 of Table 5.2.

In forward substitution, vector **v** in Eq. (3.7) is computed using a column sweep scheme. The results of vector **v** are stored in place of vector **F**. In this scheme, the data stream from matrix **L** has a constant stride of one.

In backward substitution, vector **u** in Eq. (3.6) is computed using the same strategy as with forward substitution. It should be noted that the contribution of both forward and backward substitutions to the overall CPU time in the solution of systems of linear simultaneous equations is small. For the three examples to be presented in this chapter in a subsequent section their contribution is on the order of 1 to 2% of the total CPU time.

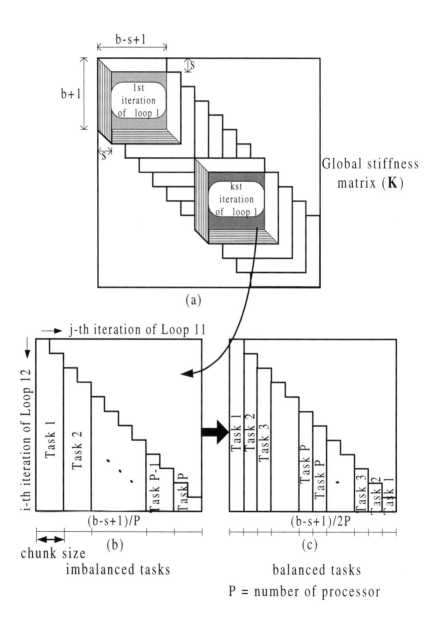

Figure 5.3: Load distribution for Cholesky factorization:
(a) Global stiffness matrix, (b) Imbalanced tasks,
and (c) Balanced tasks

5.5 CONCURRENT PRECONDITIONED CONJUGATE GRADIENT (PCG) METHOD

Iterative methods do not require the factorization of matrix **K** in Eq. (3.1). The solution is obtained through an approximate solution of Eq. (3.1) as discussed in Section 3.4. The parallel-vector preconditioned conjugate gradient algorithm is presented in Table 5.3. The major computational step for this algorithm is the matrix multiplication $\mathbf{Kp}^k$ in loops 5, 7, and 8. The stiffness matrix for a structure is symmetric and banded. For large high-rise building structures, for example, the semi-bandwidth of the stiffness matrix is substantially smaller than the number of equations. Therefore, the CPU time of the PCG method is reduced drastically when the matrix multiplication is performed in the banded form. Since the bandwidth of the banded global stiffness matrix is smaller at the two ends, the matrix multiplication is performed by dividing the rows into three parts as shown in Figure 5.4. In order to improve the efficiency of the parallel-vector algorithm, the first part of matrix multiplication is performed for rows 1 to b+1, the second part is performed for rows b+2 to $n-b+1$, and the third part for rows $n-b$ to n.

5.6 MEASURING THE PERFORMANCE OF PARALLEL-VECTOR ALGORITHMS

The goal of parallel processing is to reduce the wall-clock time that is needed to execute the job. This is done by dividing the tasks into smaller tasks and assigning them to various processors so that the tasks are performed concurrently. On the other hand, the sum of CPU times used by all processors

Table 5.3 Parallel-vector algorithm for solution of linear equations using preconditioned conjugate gradient method

Loop 1 i = 1 until n (*vectorization*)
  $r_i = f_i$
  $r_i' = f_i / K_{ii}$
  $p_i = r_i'$
Endloop 1
  Loop 2 iter = 1 until maximum iteration
    Loop 3 k = 1 until n (*vectorization*)
      $C1 = C1 + r_k' * r_k$
    Endloop 3
    Loop 4 i = 1 until b+1 (*microtasking*)
      Loop 5 j = 1 until min (i+b,n) (*vectorization*)
        $T_i = T_i + K_{ji} * p_j$
      Endloop 5
    Endloop 4
    m = 1
    ma = 2 * b
      Loop 6 ia = b+2 until n – b (*microtasking*)
        m = m + 1
        Loop 7 ja = m until m+ma (*vectorization*)
          $T_{ia} = T_{ia} + K_{ja,ia} * p_{ja}$
        Endloop 7
      Endloop 6
    Loop 8 ib = n – b +1 until n (*microtasking*)
      m = m +1
      Loop 9 jb = m until n (*vectorization*)
        $T_{ib} = T_{ib} + K_{jb,ib} * p_{jb}$
      Endloop 9
    Endloop 8

Table 5.3-continued

Loop 10 ic = 1 until n (*vectorization*)
$$C2 = C2 + P_{ic} * T_{ic}$$
Endloop 10
$\alpha = C1/C2$
Loop 11 id = 1 until n (*vectorization*)
$$x_{id} = x_{id} + \alpha * p_{id}$$
Endloop 11
Loop 12 ie = 1 until n (*vectorization*)
$$r_{ie} = r_{ie} - \alpha * T_{ie}$$
Endloop 12
Test for convergency
Loop 13 if = 1 until n (*vectorization*)
$$r'_{if} = r_{if} / K_{if,if}$$
Endloop 13
Loop 14 ig = 1 until n (*vectorization*)
$$C3 = C3 + r_{ig} + r'_{ig}$$
Endloop 14
$\beta = C3/C1$
Loop 15 = ih = 1 until n (*vectorization*)
$$p_{ih} = r'_{ih} + \beta * p_{ih}$$
Endloop 15
Endloop 2

is longer than the CPU time if the same job is executed on a single processor. This is due to the overhead in parallelizing the job. The goal of vectorization is to reduce the CPU time of each processor by broadening the vectorization portion of the code. Through judicious vectorization, performance of loops can be increased by a factor of 10 to 20. The parallelization can

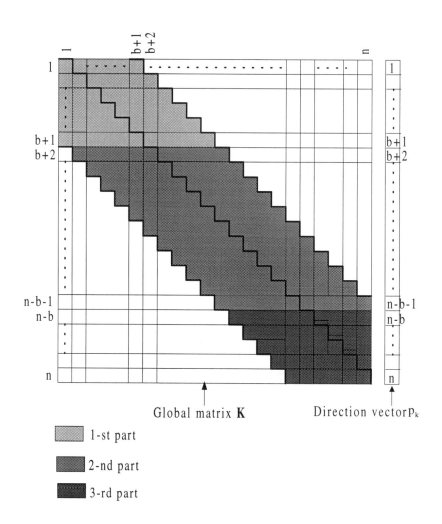

Figure 5.4: Matrix multiplication in banded form in the PCG method

reduce the wall-clock time by a maximum factor of eight on an 8-processor multiprocessor like the Cray YMP8/864. Hence, we do not want to parallelize a DO loop at the expense of vectorization. In general, the speedup of a parallel program running on p processors is defined as

$$S_p = \frac{\text{time to execute a task on one processor}}{\text{time to execute the same task on p processors running simultaneously}}$$

(5.1)

The algorithms presented in this chapter have been executed on a Cray YMP8/864, in a multi-user environment. The speedup of parallel processing is measured by using a tool called *atexpert* provided by the Cray YMP8/864 (Cray, 1991). *atexpert* estimates dedicated-machine performance based on data collected from a non-dedicated system. The speedup of vectorization is measured as the ratio of the number of floating point operations in the vectorization code to that of the non-vectorized code.

5.7 APPLICATION

The parallel-vector LU and Cholesky decomposition and preconditioned conjugate gradient algorithms have been used for analysis of three large space structures. The first example is the 848-element space truss shown in Figures 5.5a to 5.5c designed to model the exterior envelope structure of a 52-story high-rise building. The loading on the structure consists of horizontal loads acting on the exterior nodes of the space structure at every four floors (48 feet height). The horizontal loads in the y direction at each node on the sides AB and CD are obtained from the Uniform Building Code (UBC 1994) wind

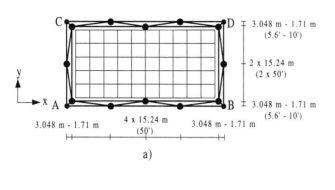

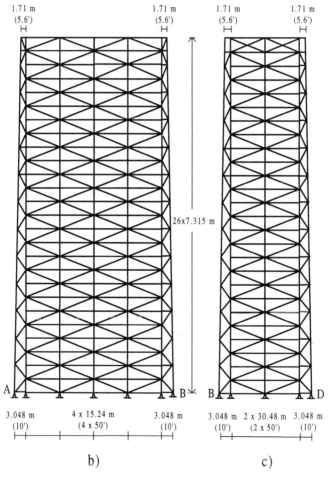

Figure 5.5: 52-story high-rise building structure: (a) Top view; (b) Front view; (c) Side view

loading using the equation $d_p = C_e C_q q_s I$, where d_p is the design wind pressure, C_e is the combined height, exposure and gust factor coefficient, C_q is the pressure coefficient, q_s is the wind stagnation pressure and I is the importance factor. The value of C_q for the inward face is 0.8 and for the leeward face is 0.5. The value of q_s is 0.6 kPa (12.6 psf) and the importance factor is assumed to be one. The values of C_s are taken from the UBC (1994), assuming exposure C (generally open area). The material is assumed to be steel with a modulus of elasticity of $1,999 \times 10^5$ kPa (29,000 ksi).

The second example is a 2756-element space truss designed to model the exterior envelope of a 162-story high-rise building (mega structure). The configuration of the structure is similar to the previous example but its height is 592.53 m (1944 feet). Its sides have the same slope. Its dimensions at the bottom are the same as those of example 1. The loading on the structure and the properties of the material are also the same as those of example 1.

The third example is the 101-story super high-rise building (mega structure), a rotated square system, consisting of 6136 elements and 1240 nodes. The perspective view of the structure is shown in Figure 5.6a and the plan is shown in Figure 5.6b. The side view is shown in Figure 5.7. The loading on the structure consists of downward vertical loads (representing the dead and live loads) and horizontal loads (representing the wind loads). The vertical loads are given as 8.90 kN (2.0 kips) at each inner node. The horizontal loads in the y direction at each node on the sides AB and DC are obtained from the Uniform Building Code (UBC, 1994). The value of coefficients for wind loading are the same as those of example 1. The material is assumed to be steel with a modulus of elasticity of $1,999 \times 10^5$ kPa (29,000 ksi).

104 Parallel-Vector Algorithms for Analysis of Large Structures

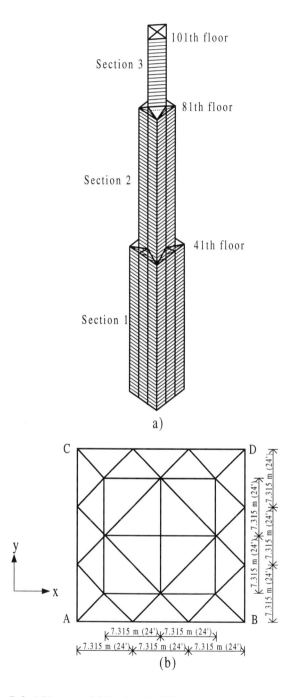

Figure 5.6: 101-story high-rise building structure: a) perspective view; b) plan

105

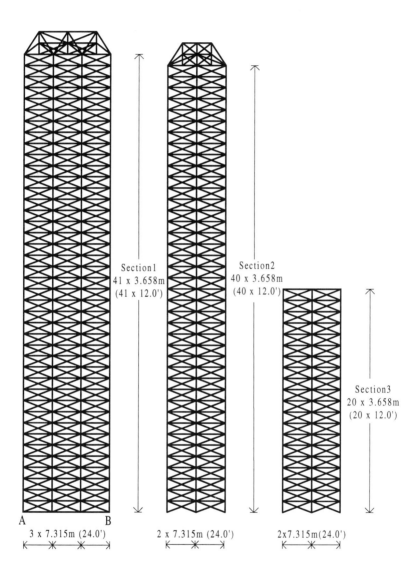

Figure 5.7: Side view of 101-story high-rise building structure

5.8 PERFORMANCE RESULTS

In this section, the performance of parallel-vector algorithms is evaluated using the three large structures presented in the previous section. The performance is measured by the speedup, CPU time, and MFLOPS on the Cray YMP8/864. Figures 5.8 to 5.10 summarize the speedups due to parallelization for assembling the global stiffness matrix and the solution of nodal displacements for the three examples using various algorithms. For evaluation and assembling element stiffness matrices into the global structure stiffness matrix, example 3 gives the best speedup. For solution of nodal displacements, the speedup of the improved Cholesky algorithm is somewhat less than those of the LU decomposition, the Cholesky, and PCG algorithms. In terms of CPU time, however, the improved Cholesky algorithm is the most efficient algorithm.

Our goal is to improve the parallel-vector Cholesky algorithm with the goal of maximizing its efficiency in terms of both vectorization and parallel processing. That is why the speedup for the improved algorithm (Table 5.2) is somewhat less than the speedup for the original algorithm (Table 5.1) and the PCG algorithm (Table 5.3). However, as presented in Table 5.4, the overall CPU time for the improved Cholesky algorithm is 23%, 24%, and 31% less than those of the original algorithm and 81%, 82%, and 96% less than those of the PCG algorithm for examples 1, 2, and 3, respectively. This saving in the CPU time increases with the size and the bandwidth of the structure. The bandwidth for the three high-rise building structures is in the range of 114 to 145. In this range, the improved parallel-vector Cholesky algorithm is the most efficient algorithm.

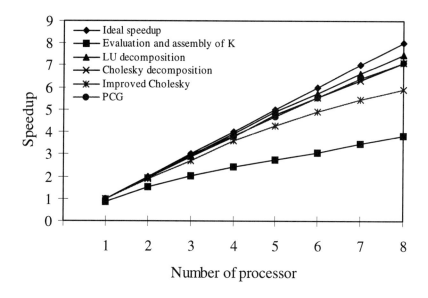

Figure 5.8: Speedup results for example 1

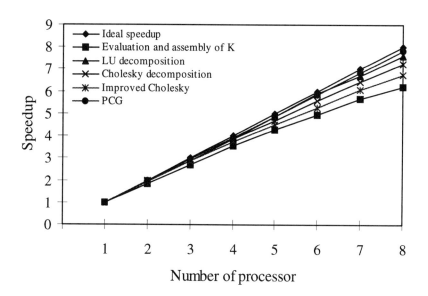

Figure 5.9: Speedup results for example 2

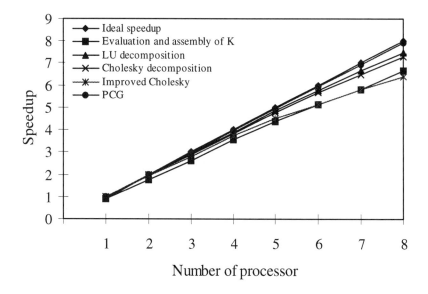

Figure 5.10: Speedup results for example 3

In order to examine the effect of the bandwith on the overall efficiency of the three parallel-vector algorithms, we used example 2 but varied the node numbering scheme in order to obtain variable bandwidths. Figure 5.11 shows the CPU time for example 2 for various semi-bandwidths. It is observed that the parallel-vector PCG algorithm becomes more efficient than the Cholesky method for a bandwith of larger than 850. For most framed structures, however, the bandwith is smaller than this value, and therefore the improved Cholesky parallel-vector algorithm is the most efficient one. For the three examples, the MFLOPS range from 168.2 to 170.1 for the LU decomposition method, from 141.50 to 144.7 for the Cholesky method (Table 5.1), from 178.2 to 207.6 for the improved Cholesky method (Table 5.2) and from 147.3 to 166.8 for the PCG approach.

Table 5.4: Performance of parallel-vector LU, Cholesky, Improved Cholesky, and PCG Algorithms

Example	Number of elements	NDOF	Semi-bandwidth	Methods	Speedup due to vectorization	MFLOPS	CPU time (s)
1	848	672	114	LU	16.64	168.2	0.1
1	848	672	114	Cholesky	16.50	141.5	0.07
1	848	672	114	Imp. Cholesky	12.28	178.2	0.06
1	848	672	114	PCG	6.68	147.3	0.29
2	2756	1968	114	LU	19.10	169.7	0.32
2	2756	1968	114	Cholesky	17.27	142.5	0.21
2	2756	1968	114	Imp. Cholesky	12.82	199.5	0.16
2	2756	1968	114	PCG	8.02	156.1	2.05
3	6136	3720	145	LU	19.24	170.1	1.01
3	6136	3720	145	Cholesky	16.51	144.7	0.61
3	6136	3720	145	Imp. Cholesky	13.27	207.6	0.42
3	6136	3720	145	PCG	8.91	166.8	9.9

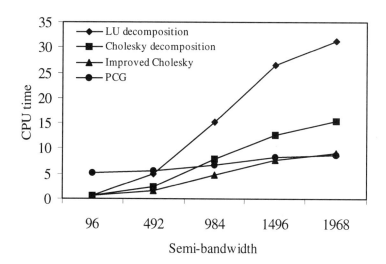

Figure 5.11: Performance of three methods as function of semi-bandwidth for example 2

CHAPTER 6
Impact of Vectorization on Large-Scale Structural Optimization

6.1 INTRODUCTION

Optimization algorithms are particularly attractive for the design of large structures with several hundreds or thousands members, such as high-rise buildings and space stations. The savings in weight (or cost) resulting from the application of these algorithms can be substantial. On the other hand, optimization of large structures subjected to realistic design constraints requires an inordinate amount of computer processing resources. With the availability of the new generation multiprocessor supercomputers, optimization of very large and complex structures is achievable. The challenge is to develop special algorithms utilizing the unique architecture of these machines.

Most of the computer processing time in structural optimization algorithms, such as the optimality criteria approach, is spent in iterative arithmetic operations (DO loops). Thus, computer processing time can be minimized by broadening the vectorizable portion of the code. In this chapter, the impact of vectorization in the inner DO loops on the performance of structural optimization algorithms is discussed.

6.2 OPTIMALITY CRITERIA APPROACH

Two fundamentally different approaches have been commonly used in structural optimization with the objective of minimizing the weight of the structure subjected to constraints. Both methods are non-linear and have iterative solutions. The first approach is called Mathematical Programming (MP). Quadratic Programming, Sequential Linear Programming, and Steepest Descent are among the MP methods. The selection of the initial design plays an important role on the rate of convergence of these methods. Poor initial design may lead to prohibitively high computational cost. In general, the search step in the MP methods can be formulated as follow:

$$\mathbf{X}_{p+1} = \mathbf{X}_p + \alpha_p \mathbf{S}_p \tag{6.1}$$

where $\mathbf{X}$ is the vector of design variables, α is the step size, $\mathbf{S}$ is the search direction vector, and p is the iteration number.

The second popular approach to structural optimization is the Optimality Criteria (OC) approach which is based on an approximation concept utilizing knowledge of structural behavior. In the OC method, recurrence relations are derived based on displacements, stress and frequency constraints. In addition to the recurrence relations, a scaling strategy to reach the surface of the feasible domain has to be devised. After computing the design variables based on the recurrence relations, the new design variables for displacement constraints are found from

$$\mathbf{X}_{p+1} = \Omega \mathbf{X}_p \qquad (6.2)$$

where Ω is the scaling factor or a distance to move along the direction vector $\mathbf{X}_p$.

The behavior of an MP approach such as Steepest Descent method is compared with an OC approach in Figure 6.1. This figure represents a two-dimensional design space with variables x_1 and x_2 and boundaries of constraints shown with curves g_1 and g_2. Both approaches require the evaluation of constraint gradients which is referred to as sensitivity analysis. In the evaluation of constraint gradient, the solution of the systems of simultaneous linear equations is involved. The iteration paths of the MP and OC methods are shown by solid and dashed lines in Figure 6.1. In each iteration, the constraint gradient has to be evaluated, and this is the most time-consuming step in the iteration process. The number of iterations affects the computing time and consequently affects the choice of the optimization method.

6.3 SENSITIVITY ANALYSIS

Computation of the gradients or derivatives of the constraints in the gradient-based optimization techniques is time consuming and impacts the overall efficiency significantly. As such, efficient evaluation of the gradients is of primary concern in developing optimization algorithms. The two fundamental approaches used to evaluate the gradients of constraints are the analytical approach and the finite difference method. An analytical approach based on the virtual load method is

114 Impact of Vectorization on Large-Scale Structural Optimization

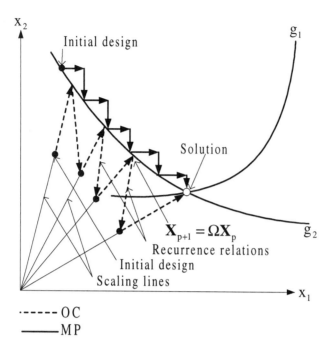

Figure 6.1: The steps of Mathematical Programming and Optimality Criteria methods

presented in this section.

Consider the equilibrium equation

$$\mathbf{K}\mathbf{u}_k = \mathbf{F}_k \tag{6.3}$$

where $\mathbf{F}_k$ and $\mathbf{u}_k$ are the load and displacement vectors due to loading condition k, respectively. The displacement at jth degree of freedom due to loading condition k can be written as follows:

$$u_{jk} = \mathbf{F}_i^T \mathbf{u}_k \tag{6.4}$$

where $\mathbf{F}_i^T$ is the transpose of the virtual load vector. The entry of load vector $\mathbf{F}_i^T$ is equal to one for i = j and zeros elsewhere. Eq. (6.4) in expanded form can be written as follows:

$$u_{jk} = [F_1 \ F_2 \ \ldots \ F_i \ \ldots \ F_n] \begin{bmatrix} u_{1k} \\ u_{2k} \\ .. \\ u_{jk} \\ .. \\ u_{nk} \end{bmatrix} \quad (6.5)$$

Differentiating Eq. (6.4) with respect to design variable A_m (cross section of member m), we obtain

$$\frac{\partial u_{jk}}{\partial A_m} = \frac{\partial F_j^T}{\partial A_m} u_k + F_j^T \frac{\partial u_k}{\partial A_m} \quad (6.6)$$

Differentiating Eq. (6.3) with respect to design variable A_m, we obtain

$$\frac{\partial F_k}{\partial A_m} = K \frac{\partial u_k}{\partial A_m} + \frac{\partial K}{\partial A_m} u_k \quad (6.7)$$

In most structural problems, the loads are not a function of design variables. Therefore, the derivative of F_k with respect to A_m is equal to zero, and Eq. (6.7) becomes

$$\frac{\partial u_k}{\partial A_m} = -K^{-1} \frac{\partial K}{\partial A_m} u_k \quad (6.8)$$

In order to calculate derivatives of the displacement at the jth degree of freedom with respect to the design variables, the relation between the virtual load and displacement is expressed as follows:

$$K v_j = F_j \quad (6.9)$$

where F_j and v_j are the virtual load and displacement vectors. The elements of F_j are zeros except the jth term which is equal to one. Transposing and rearranging Eq. (6.9) yields

$$v_j^T = F_j^T K^{-1} \tag{6.10}$$

Substituting Eqs. (6.8) and (6.10) into Eq. (6.6) and noting that matrix K is symmetric, we obtain

$$\frac{\partial u_{jk}}{\partial A_m} = -v_j^T \frac{\partial K}{\partial A_m} u_k \tag{6.11}$$

In computing the derivative of the global stiffness matrix ($\partial K / \partial A_m$), only the stiffness matrix of member m contributes to the gradient. Since the member stiffness matrix is a linear function of design variables A_m, the derivative of global stiffness matrix K with respect to design variables can be written as follows:

$$\frac{\partial K}{\partial A_m} = \frac{k_m}{A_m} \tag{6.12}$$

where k_m is the stiffness matrix of member m. Substituting Eq. (6.12) into Eq. (6.11), we obtain the displacement constraint gradients

$$\frac{\partial u_{jk}}{\partial A_m} = -v_{mj}^T \frac{k_m}{A_m} u_{mk} \tag{6.13}$$

where u_{mk} is the displacement vector of member m due to the loading condition k, and v_{mj} is the virtual displacement vector of member m due to the application of a unit load in the direction of the jth degree of freedom.

6.4 RECURRENCE RELATIONS FOR TRUSS STRUCTURES

In this section the OC method is chosen for formulating the structural optimization problem. This method is not as general as MP methods. The primary reason to choose this method is its simplicity and relative efficiency for large structures. Lagrange multipliers are used to minimize the weight of the structure subjected to displacements and stress constraints. Displacement gradients are evaluated using the virtual load method described in the previous section. Displacement constraints are divided into active constraints and inactive constraints. A displacement constraint is called active when at a particular point and direction, it is in the neighborhood of the limiting value within a given tolerance. In the derivation of the recurrence relation only active constraints are considered for evaluating the Lagrange multipliers.

The structural optimization problem can be stated as follows: Find the vector of design variables (cross-sectional areas) such that the weight of the structure

$$W = \sum_{m=1}^{M} \rho_m L_m A_m \tag{6.14}$$

is minimized subject to the following stress, displacement, and fabricational constraints:

$$\sigma_m^L \leq \sigma_{mk} \leq \sigma_m^U, \quad m = 1, \ldots, M, \, k = 1, \ldots, L \tag{6.15}$$

$$|u_{jk}| \leq r_j, \, j = 1, \ldots, N, \, k = 1, \ldots, L \tag{6.16}$$

$$A_m^L \leq A_m \leq A_m^U \tag{6.17}$$

where W is the total weight of the structure, k is the loading number, M is the number of members, N is the number of degrees of freedom, L is the number of loading conditions, ρ_m is the unit weight of member m, L_m is the length of member m, A_m is the cross-sectional area of member m, A_m^L and A_m^U are the lower and upper bounds on the cross-sectional area, respectively, σ_m^L and σ_m^U are the lower and upper bounds on the stress constraint of member m, σ_{mk} is the stress in member m due to loading condition k, and r_j is the maximum displacement at the jth degree of freedom.

Since high-rise building structures are often controlled by displacement constraints, the derivation of the optimality condition is based on displacement constraints. Those members that violate the stress constraints are scaled at the members' level.

The Lagrangian fucntion is defined as follows:

$$L(\lambda_{jk}, A_m) = \sum_{m=1}^{M} \rho_m L_m A_m + \sum_{j=1}^{N}\sum_{k=1}^{L} \lambda_{jk}(u_{jk} - r_j) \qquad (6.18)$$

where λ_{jk} is the Lagrange multiplier for the displacement constraint associated with the jth constrained degree of freedom and the kth loading case. When only active displacement constraints are considered, we have

$$L(\lambda_{jk}, A_m) = \sum_{m=1}^{M} \rho_m L_m A_m + \sum_{j=1}^{N_{ac}}\sum_{k=1}^{L} \lambda_{jk}(u_{jk} - r_j) \qquad (6.19)$$

where N_{ac} is the number of active displacement constraints.

In order to obtain the optimality condition, the following Kuhn-Tucker conditions are satisfied: The derivatives of the

Lagragian function with respect to design variable A_m must be equal to zero.

$$\rho_m L_m + \sum_{j=1}^{N_{ac}} \sum_{k=1}^{L} \lambda_{jk} \frac{\partial u_{jk}}{\partial A_m} = 0 \qquad (6.20)$$

where

$$\lambda_{jk} \geq 0, \quad j=1,\ldots\ldots,N_{ac} \qquad (6.21)$$

Substituting Eq. (6.13) into Eq. (6.20) yields

$$\rho_m L_m = \sum_{j=1}^{N_{ac}} \sum_{k=1}^{L} \lambda_{jk} v_{mj}^T \frac{k_m}{A_m} u_{mk}, m = 1, \ldots\ldots, M \qquad (6.22)$$

or

$$A_m \rho_m L_m = \sum_{j=1}^{N_{ac}} \sum_{k=1}^{L} \lambda_{jk} v_{mj}^T k_m u_{mk}, m = 1, \ldots\ldots, M \qquad (6.23)$$

For the jth active displacement constraint and the kth loading condition, Eq. (6.23) becomes

$$A_m \rho_m L_m = \lambda_{jk} v_{mj}^T k_m u_{mk}, \quad m = 1, \ldots\ldots, M \qquad (6.24)$$

or

$$\sum_{m=1}^{M} A_m \rho_m L_m = \lambda_{jk} \sum_{m=1}^{M} v_{mj}^T k_m u_{mk} \qquad (6.25)$$

From Eqs. (6.4) and (6.9), we obtain

$$u_{jk} = v_j^T K u_k \qquad (6.26)$$

or

$$u_{jk} = \sum_{i=1}^{M} v_{mj}^T k_m u_{mk} \qquad (6.27)$$

From Eqs. (6.25) and Eq. (6.27), we obtain

$$W = \lambda_{jk} u_{jk} \tag{6.28}$$

The Lagrange multipliers for the kth loading condition can be obtained from Eq. (6.28):

$$\lambda_{jk} = \frac{W}{u_{jk}} \tag{6.29}$$

Substituting Eq. (6.29) into Eq. (6.24), we obtain

$$A_m \rho_m L_m = \frac{W}{u_{jk}} \mathbf{v}_{mj}^T \mathbf{k}_m \mathbf{u}_{mk}, \quad m = 1, \ldots, M \tag{6.30}$$

From Eq. (6.30) the optimality condition can be written as follows:

$$\left(\frac{W}{u_{jk}}\right)\left(\frac{\mathbf{v}_{mj}^T \mathbf{k}_m \mathbf{u}_{mk}}{\rho_m L_m A_m}\right) = 1, \quad m = 1, 2, \ldots M \tag{6.31}$$

When there is only one active displacement constraint, the following recurrence relationship is used (Khot and Berke, 1984)

$$(A_m)_{p+1} = (A_m)_p \left\{ \left[\lambda_{jk} \left(\frac{\mathbf{v}_{mj}^T \mathbf{k}_m \mathbf{u}_{mk}}{\rho_m L_m A_m} \right) \right]^{\varsigma} \right\}_p, \quad m = 1, 2, \ldots M \tag{6.32}$$

where p is the iteration number and ς determines the step size.

The parameter ς controls the convergence and stability of the method. In this work, we start with $\varsigma = 0.5$. When oscillation occurs in the objective function at some iteration, the value of ς is reduced to $\varsigma/2$ before proceeding to the next iteration.

When more than one constraint is active, the Langrange multipliers are computed from

$$(\lambda_{jk})_{p+1} = (\lambda_{jk})_p \left\{ \left(\frac{u_{jk}}{r_j} \right)^{\varphi} \right\} \tag{6.33}$$

where φ determines the step size (a value of $\varphi = 2$ is used in this work), and the following recurrence relationship is used:

$$(A_m)_{p+1} = (A_m)_p \left\{ \sum_{c=1}^{N_{ac}} \left[(\lambda_{jk})_c \left(\frac{v_{mj}^T k_m u_{mk}}{\rho_m L_m A_m} \right)_c \right] \right\}_p^{\varsigma}, \quad m=1,2,...M \tag{6.34}$$

where Nac is the number of active displacement constraints.

When the stress constraints are violated, the cross-sectional areas of the members are obtained from the stress ratio relationship in the following form:

$$(A_m)_{p+1} = (A_m)_p \frac{(\sigma_m)_k}{\sigma_m^U \text{ or } \sigma_m^L} \tag{6.35}$$

Eq. (6.35) is applied repeatedly until the displacement constraint controls the design, or all members are fully stressed, or a certain tolerance is achieved.

6.5 VECTORIZED OPTIMIZATION ALGORITHM

The vectorized algorithm for optimization of space axial-load structures is presented in Table 6.1. In the structural analysis problem, the displacement vector is obtained by solving a linear system of equilibrium equations in the following form:

Table 6.1 Vectorized Algorithm for Optimization of Space Axial-Load Structures

Step 1
Read in the input data. Choose initial cross-sectional areas, tolerance, and step size parameters.

Step 2
Assemble the structure stiffness matrix
    For member i = 1 until M (vectorization)
        *Calculate member stiffness matrix **k** (6,6,i)*
    Next member
    For member i = 1 until M
        Assemble the member stiffness matrix into the global stiffness matrix

$$\mathbf{K} = \sum_{i=1}^{M} \mathbf{k}(6,6,i)$$

    Next member
    For nodal degree of freedom j = 1 until NP (*vectorization*)
        Assemble the nodal point loads using loop unrolling vertically

$$\mathbf{F} = \sum_{j=1}^{NP} F_j$$

    Next nodal degree of freedom.

Step 3
Solve for nodal displacements
Phase = 1 when the new iteration starts or stress constraints govern.
Phase = 2 when max $|r_{jk} - r_j| >$ tolerance

If phase = 1, solve for the nodal displacements (Table 6.2)
If phase = 2, continue
For loading condition k

Table 6.1-continued

For member i = 1 until M (*vectorization*)
$\quad A_i = (SFD)_k * A_i$
Next member
For loading condition k
$\quad$ For nodal degree of freedom j = 1 until N (*vectorization*)
$\quad\quad u_{jk} = (SFD)_k * u_{jk}$
$\quad$ Next nodal degree of freedom
For loading condition k
For j = 1 until N
$\quad$ For m = 1 until N
$\quad\quad K_{mj} = K_{mj} * (SDF)_k$
$\quad$ Next m
Next j

Step 4

Find member forces and stresses for each loading condition
If phase = 1, evaluate member forces and stresses
For loading condition k = 1 until L
$\quad$ For member i = 1 until M (*vectorization*)
$\quad\quad \sigma_{ik} = \delta_{ik} \dfrac{E}{L_i}$
$\quad\quad$ (δ_{ik} *is the axial deformation of member i due to the kth loading condition*)
$\quad$ Next member
Next loading

Step 5

Find the maximum scaling factor
For loading condition k = 1 until L
$\quad$ For nodal degree of freedom i = 1 until N
$\quad\quad$ *Find the maximum scaling factor for displacement* $(SFD)_k$

Table 6.1-continued

Next nodal degree of freedom
For member i = 1 until M
   Find the maximum scaling factor for stress $(SFS)_k$
Next member
Next loading

Step 6
Scale the cross-sectional areas
Find the maximum of $(SFD)_k$ and $(SFS)_k$
If $(SFD)_k > (SFS)_k$, find Dif = |SFD − 1.0|

  If Dif > Tolerance,
    Scale the stiffness matrix
      For j = 1 until N
        For m = 1 until N (*vectorization*)
$$K_{mj} = K_{mj} * (SDF)_k$$
        Next m
      Next j
    Scale the nodal displacement vector
      For nodal degree of freedom j = 1 until N (*vectorization*)
$$u_{jk} = u_{jk} * (SFD)_k$$
      Next nodal degree of freedom
    Scale the design variables and member stresses
      For member i = 1 until M (*vectorization*)
$$(A_i)^{new} = (A_i)^{old} * (SDF)_k$$
      Next member
      For member i = 1 until M (*vectorization*)
$$\sigma_{ik} = \sigma_{ik} * (SFD)_k$$
      Next member
      Go to step 7
Else

Table 6.1-continued

If SFS > SFD or phase = 1,
   Scale the design variables
      For member i = 1 unil M (*vectorization*)

$$(A_i)^{new} = (A_i)^{old} \left[\frac{\sigma_{ik}}{\sigma_i^L \text{ or } \sigma_i^u} \right]$$

Next member
Go to step 2

Step 7
Calculate the weight of the structure
For member *i* = 1 until M (*vectorization*)

$$W = \sum_{i=1}^{M} A_i * L_i * \rho_i$$

Next member

Step 8
Find the active constraints and calculate the Lagrange multipliers

If one constraint is active, calculate the Lagrange multiplier from

$$\lambda_{jk} = \frac{W}{r_j}$$

If more than one constraint is active, the Lagrange multipliers are computed from

$$(\lambda_{jk})_{p+1} = (\lambda_{jk})_p \left\{ \left(\frac{u_{jk}}{r_j} \right)^{\varphi} \right\}_p$$

Step 9
Calculate the virtual nodal displacements

Table 6.1-continued

Apply a unit load in the direction of the active displacement constraint and evaluate the virtual nodal displacements, v_{ij}

Step 10
Calculate displacement gradients
For member i = 1 until M (*vectorization*)
   Calculate member stiffness matrix **k**(6,6,i)
Next member
For loading condition k = 1 until L
   For member i = 1 until M (*vectorization*)
     Calculate member displacements vector $\mathbf{u}_{ik}$
   Next member
   For member i = 1 until M (*vectorization*)
     Calculate member virtual displacements vector $\mathbf{v}_{ij}$
   Next member
Next loading

Step 11
Calculate the displacement gradients
For member i = 1 until number of members (*vectorization*)
   Calculate the displacement gradients

$$g_{ijk} = -\mathbf{v}_{ij}^T \frac{\mathbf{k}_i}{A_i} \mathbf{u}_{ik} \text{ using unrolling loop vertically}$$

Next member

Step 12
Calculate the optimality condition coefficient
For member i = 1 until M (*vectorization*)

$$C_i = \lambda_{jk} \frac{g_{ijk}}{L_i \rho_i}$$

Next member

Table 6.1-continued

Step 13
Find the new design variables
For member i = 1 until M *(vectorization)*
  Calculate the new design variables

$$(A_i)_{p+1} = (A_i)_p \left\{ \left[(\lambda_{jk}) \left(\frac{\mathbf{v}_{ij}^T \mathbf{k}_i \mathbf{u}_{ik}}{\rho_i L_i A_i} \right) \right]^\varsigma \right\}_p$$

(for one active constraint)

$$(A_i)_{p+1} = (A_i)_p \left\{ \left[\sum_{c=1}^{N_{ac}} (\lambda_{jk})_c \left(\frac{\mathbf{v}_{ij}^T \mathbf{k}_i \mathbf{u}_{ik}}{\rho_i L_i A_i} \right)_c \right]^\varsigma \right\}_p$$

(for more than one active constraint)
Next member

Step 14
Go to step 2 to resume the next iteration

$$\mathbf{K}\mathbf{u} = \mathbf{F} \qquad (6.36)$$

$$\mathbf{K} = \sum_{m=1}^{M} \mathbf{k}_m \text{ and } \mathbf{F} = \sum_{j=1}^{NP} \mathbf{F}_j \qquad (6.37)$$

where **K** is the global stiffness matrix and **F** is the global load vector, NP is the total number of nodal points and $\mathbf{F}_j$ is the nodal load vector at the jth nodal point.

First, member stiffness matrices are evaluated one by one. Next, they are assembled into the global stiffness matrix. This way, dependency relationships are avoided. However, we need to enlarge the dimension of the member stiffness matrix from **k**(6,6) to **k**(6,6M) (for space axial-load structures). The evaluation of member stiffness matrices can now be vectorized.

The assembly of the nodal point loads to obtain the global load vector can be performed by the following scheme:

>**Loop 1** i = 1,NP
> **Loop 2** j = 1,DOF
>   Assemble the nodal point loads
> **Endloop 2**
>**Endloop 1**

where DOF is the number of degrees of freedom per node. In this scheme, only the inner loop j is vectorizable, which is a short DO loop. The performance of this operation can be improved by applying loop unrolling, that is to remove the inner DO loop by repeating the statements in that loop, so that the outer DO loop can be vectorized.

Solution of linear simultaneous equations is the most time-consuming part in the structural optimization of large structures. We use the LU and Cholesky decomposition methods for solving linear simultaneous equations. For the sake of brevity, we present only the vectorized algorithm for the solution of linear equations using the Cholesky decomposition approach in Table 6.2. These methods consist of three steps: factorization, forward substitution, and backward substitution. For large structures, most of the computational time is spent in the first step, that is, factorization. There are no dependencies in the inner DO loops of this algorithm. Therefore, all inner DO loops are vectorized.

LU and Cholesky factorization methods have advantages for solving linear simultaneous equations in the optimality criteria approach. First, when the displacement constraint governs, the nodal displacements due to virtual load are

Table 6.2 Vectorized algorithm for solution of linear equations using the banded storage Cholesky decomposition

1. *Store the stiffness matrix K with dimensions $n \times n$, load vector F with dimensions $n \times 1$, displacements vector u with dimensions $n \times 1$, and the semi-bandwidth b*

2. *Transform the stiffness matrix to lower triangular matrix*
    Loop 1 k = 1 until n − 1
    $$K_{kk} = K_{kk}^{1/2}$$
    Loop 2 s = k+1 until min (k+b,n) (*vectorization*)
    $$K_{sk} = K_{sk} / K_{kk}$$
    Endloop 2
    Loop 3 j = k+1 until min (k+b,n)
        Loop 4 i = j until min (k+b,n) (*vectorization*)
        $$K_{ij} = K_{ij} - K_{ik} * K_{jk}$$
        Endloop 4
    Endloop 3
    Endloop 1
    $$K_{nn} = K_{nn}^{1/2}$$

3. *Forward substitution*
    Compute vector y and store it in F ($L\,y = F$)
    Loop 1 j = 1 until n
    $$F_j = F_j / K_{jj}$$
    Loop 2 i = j+1 until min (j+b,n) (*vectorization*)
    $$F_i = F_i - K_{ij} * F_j$$
    Endloop 2
    Endloop 1

Table 6.2-continued

4. Backward substitution
*Compute the displacement vector **u** from the lower triangular matrix and store it in **F** (**K u** = **F**)*
    Loop 1 j = n until 1
        $F_j = F_j / K_{jj}$
        Loop 2 i = j−1 until max (j−b,1) (*vectorization*)
            $F_i = F_i - K_{ji} * F_j$
    Endloop 2
    Endloop 1

obtained from forward and backward substitutions only. Second, when the displacement constraint is violated, the nodal displacements are scaled, and the factorized stiffness matrix is obtained by simply scaling the previously factorized stiffness matrix. The strategy reduces the computational time significantly. The performance of these two algorithms in solving the structural optimization problems will be discussed in a subsequent section.

6.6 APPLICATION

Example 1 72-member space truss

The 72-member space truss shown in Figure 6.2 consists of 72 members and 20 nodes. The same structure has been solved by Adeli and Kamal (1986) and others. This structure is subjected to two different loading cases. In the first case, loading on the structure consists of three 5-kip forces in the x, y and z-directions at node 17. In the second case, loading on the structure consists of four 5-kip forces in the z-direction at nodes

131

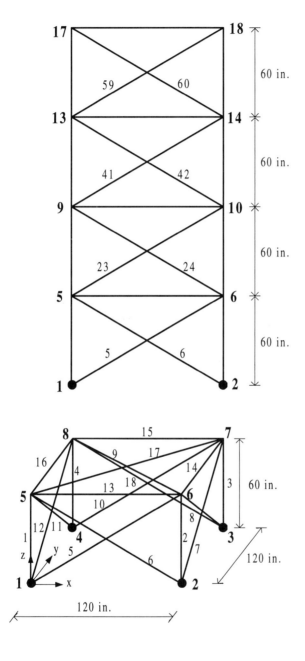

Figure 6.2: 72-member space truss

17, 18, 19, and 20. The member stresses are limited to 25 ksi in both tension and compression. The maximum displacement in the x, y, and z directions at the nodes on the top of the structure is limited to 0.25 in., modulus of elasticity of the material is 10000 ksi and the weight of the material is 0.1lb/in^3. The lower and the upper bounds of the cross-sectional areas are 0.1 in^2 and 80.0 in^2. The maximum number of iterations is set to 10. The convergence history for this example is shown in Figure 6.3. The computational processing time, number of floating point operations per second and the speedup due to vectorization using LU and Cholesky decompositions are summarized in Table 6.3.

Example 2 52-story high-rise building structure

The second example is the 848-member space truss structure shown in Figures 5.5a to 5.5c designed to model the exterior envelope structure of a 52-story high-rise building (mega structure). The loading and properties are the same as example 1 of Chapter 5. The lower and upper bounds of the cross-sectional areas in this example are 1.0 in^2 and 500 in^2. The member stresses are limited to 25.0 ksi in both tension and compression. The material is assumed to be steel with modulus of elasticity of 29,000 ksi and the unit weight of material of 0.284 lb/in^3. The displacement constraints are given as $\pm$ 18.0 in. in the y direction for the nodes on the top floor (equal to 0.0025 H, where H is the height of structure). The maximum number of iterations is set to 20. Figure 6.4 shows the convergence history. A minimum weight of 442.5 MN (99,571 kips) is found after 17 iterations. The computational processing time, number of floating point operations and speedup due to

133

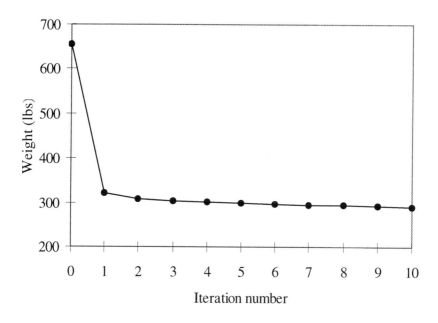

Figure 6.3: Convergence history for the 72-member space truss

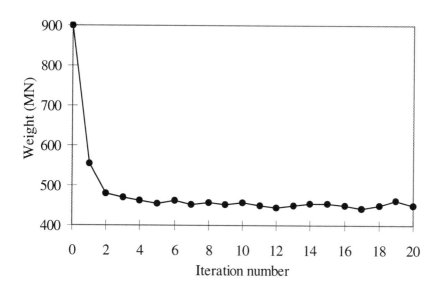

Figure 6.4: Convergence history for the 52-story high-rise building structure

vectorization using LU and Cholesky decompositions are summarized in Table 6.3.

Example 3 162-story super high-rise building structure

This example is a 2756-member space truss designed to model the exterior envelope of a 162-story super high-rise building (mega structure). The configuration of this structure is similar to the previous example but its height is 162 x 12 ft = 1944 feet. Its sides have the same slope. Its dimensions at the bottom are the same as those of example 2. The loading on the structure and the properties of the material are also the same as those of example 2. The displacement constraints are given as ± 58 in. in the y direction for the nodes on the top floor (equal to 0.0025 H). Figure 6.5 shows the convergence history. A minimum weight of 1,679 MN (377,861 kips) is obtained after 11 iterations. The computational time, number of floating point operations and the speedup due to vectorization using LU and Cholesky decompositions are summarized in Table 6.3.

Example 4 101-story super high-rise building

The fourth example is a 101-story super high-rise building shown in Figure 5.6a to 5.6c. The loading and properties are the same as example 3 of Chapter 5. The lower and upper bounds of the cross-sectional areas in this example are 1.0 in^2 and 500.0 in^2. The member stresses are limited to 25.0 ksi for both tension and compression. The displacement constraints are given as ± 36.0 in. in the y direction for the nodes on the top level (equal to 0.0025H). The material is assumed to be steel with modulus of elasticity of 29000 ksi and

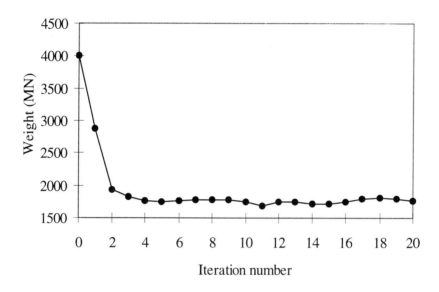

Figure 6.5: Convergence history for the 162-story high-rise building structure

the unit weight of the material 0.284 lb/in^3. The maximum number of iterations is set to 20. Figure 6.6 shows the convergence history. A minimum weight of 199.4 MN (44,862 kips) is found after 19 iterations. The computational time, the number of floating point operations and the speedup due to vectorization using LU and Cholesky decompositions are summarized in Table 6.3.

6.7 SUMMARY OF RESULTS AND CONCLUSION

We presented a vectorized algorithm for optimization of large structures in this chapter. The algorithm was applied to the minimum weight design of four structures. The first example, a small structure with 72 members and 60 degrees of freedom, was taken from the literature to verify the vectorized algorithm.

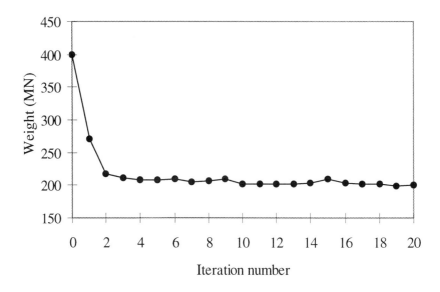

Figure 6.6: Convergence history for the 101-story high-rise building structure

The other three examples are large structures with the number of members ranging from 848 to 6136, and number of degrees of freedom ranging from 672 to 3720. No design linking strategy has been used because our goal was to perform large-scale optimization. Thus, the number of design variables in each example is equal to the number of members in that structure. Our largest example has 6136 design variables with 3720 degrees of freedom. This is a very large-scale optimization problem.

Table 6.3 summarizes the numerical results obtained and the performance of the vectorized structural optimization algorithm for the four examples. The algorithm has been implemented in FORTRAN using the Cray CF77 compiler. Performance is measured by MFLOPS and the CPU time on the

Table 6.3: Performance of vectorized algorithm for optimization of space axial-load structures

Member No.		Example 1	Example 2	Example 3	Example 4
No. of Nodes		20	224	656	1240
NDOF		60	672	1968	3720
No. of members		72	848	2756	6136
MFLOPS with vectorization	Cholesky	22.30	129.90	129.00	124.60
	LU	26.10	162.60	162.30	154.80
MFLOPS without vectorization	Cholesky	7.60	8.50	8.50	8.50
	LU	7.60	8.60	8.60	8.60
CPU time with vectorization (sec.)	Cholesky	0.10	27.60	129.10	184.40
	LU	0.10	42.70	200.70	269.80
CPU time without vectorization (sec.)	Cholesky	0.20	442.40	2076.90	2729.10
	LU	0.30	808.90	3905.30	4829.40
Speedup due to vectorization	Cholesky	2.00	16.00	16.10	14.80
	LU	3.00	18.90	19.50	17.90

Cray YMP8/864 for both Cholesky and LU decompositions. In addition, we measure speedup as the ratio of the CPU time of the non-vectorized code to that of the corresponding vectorized code.

For all four examples, LU decomposition provides a higher MFLOPS and speedup. For the small example one, maximum values of 26.1 and 3.0 were obtained for the MFLOPS and speedup, respectively. These values are not high because the length of vectors in this example is small and vectorization capability cannot be fully utilized. For the other three large example, however, MFLOPS ranges from 154.8 to 162.6, and the speedup ranges from 14.8 to 19.5. This variations stems partly from constraint conditions of the problems. When a stress constraint becomes active, the corresponding scaling operations impede the vector chaining, resulting in the reduction of MFLOPS.

On the other hand, using the Cholesky decomposition reduces the total CPU time for all four examples, and substantially so for large structures. In this case, speedup due to vectorization is 2.0 for the small example one, and ranges from 14.8 to 16.0 for the three large structures. Thus, despite the fact that we found the LU decomposition to be more suitable for vectorization than the Cholesky decomposition, the additional gain is not large enough to compensate for additional computations involved in the LU decomposition. Still, it should be noted that even with the Cholesky decomposition we achieved a significant speedup of 14.8 to 16.0 for large structures.

In conclusion, this work demonstrates that we can increase the speedup of an optimization algorithm significantly through the adroit and judicious use of vectorization techniques. Additional speedup can be achieved through the use of parallel processing techniques (Adeli, 1992a&b). But this work shows that on coarse-grain multiprocessors (those with only a few processors) the impact of vectorization can be substantially more than that of parallel processing. For example, the maximum theoretical speedup of the parallel algorithm on an eight-processor machine such as Cray YMP8/864 or Encore Multimax is eight (Adeli and Kamal, 1992a&b; Hsu and Adeli, 1991). In this investigation, however, we achieved a speedup of at least 14.8 for large structures due to vectorization only.

CHAPTER 7
Optimization of Large Steel Structures Using Standard Cross Sections

7.1 INTRODUCTION

A number of papers have been published recently on the subject of optimization of large structures assuming continuous design variables (for example, Hsu and Adeli, 1991; Adeli and Cheng, 1993, 1994; Soegiarso and Adeli, 1994). This chapter is concerned with the optimization of large steel structures consisting of a few thousand members using standard cross sections such as wide flange (W) shapes included in the American Institute of Steel Construction (AISC) manuals (AISC, 1989, 1994). A multi-constraint optimality criteria discrete optimization algorithm is presented for minimum weight design of structures subjected to stress, displacement, and buckling constraints. The structures are subjected to the actual constraints of AISC Allowable Stress Design (ASD) and Load and Resistance Factor Design (LRFD) specifications (AISC, 1989, 1994). A design linking strategy is used to reduce the number of design variables and incorporate the fact that, in practice, the same shape is used for members with similar physical and loading conditions.

Cross-sectional areas of members are normally used as design variables. When buckling constraints are considered, two additional difficulties are encountered. First, the radius of gyration of the cross section appears in the buckling constraint. In order to avoid doubling the number of design variables, various researchers have tried to relate the radius of gyration to the cross-sectional area approximately (for example, Adeli and Balasubramanyam, 1988; John and Ramakrishnan, 1990; Chan, 1992), Grierson and Chan, 1993). Second, buckling constraints are nonlinear and implicit functions of design variables. This can create convergence difficulties.

Our discrete optimization algorithm includes an efficient integer mapping strategy for mapping the computed cross-sectional areas to the available W shapes. This strategy includes the creation of a tree network and an integer search method.

The algorithm has been applied to the minimum weight design of five high-rise and super high-rise structures ranging in height from 41 stories to 81 stories and in size from 848 to 5860 members.

7.2 AN OPTIMALITY CRITERIA APPROACH

In this section, we present an optimality criteria approach for optimization of steel structures subjected to stress, displacement, and buckling constraints based on the AISC ASD or LRFD specifications. The algorithm is general, but we limit the scope of this chapter to axial-load space structures. The general structural optimization problem with the design linking strategy can be stated in the following form: Find the set of design variables, A_i (cross-sectional areas), such that the weight of the structure

$$W = \sum_{i=1}^{N_d} \rho_i A_i \sum_{m=1}^{N_{mi}} L_{im} \qquad (7.1)$$

is minimized subject to the displacement, fabrication, stress, and buckling constraints to be described shortly. In Eq. (7.1), N_d is the number of design variables (groups of members with identical cross sections), L_{im} is the length of member m belonging to group i, ρ_i is the unit weight of members in group i, N_{mi} is the number of members in group i, and A_i is the cross-sectional area of members in group i.

The displacement and fabrication constraints are

$$r_j^L \leq u_{jk} \leq r_j^U, \quad j = 1,\ldots\ldots,N \quad k = 1,\ldots\ldots,L \qquad (7.2)$$

$$A_i^L \leq A_i \leq A_i^U, \quad i = 1,\ldots\ldots,N_d \qquad (7.3)$$

where N is the total number of displacement degrees of freedom, L is the number of loading conditions, A_i^L and A_i^U are the lower and the upper bounds on the cross-sectional area of members in group i, u_{jk} is the displacement of the jth degree of freedom due to loading condition k, and r_j^L and r_j^U are the lower and upper bounds on the displacement of the jth degree of freedom.

A. Stress and buckling constraints based on the AISC ASD specifications

Using positive values for tensile stresses and negative values for compressive stresses, the stress constraints according to the AISC ASD specifications can be expressed as

$$-F_{am} \leq \sigma_{mk} \leq 0.6F_y, \quad m = 1,\ldots,N_m \qquad (7.4)$$

where σ_{mk} is the stress in member m due to the loading condition k, F_y is the yield stress of steel, N_m is the total number of members in the structure equal to

$$N_m = \sum_{i=1}^{N_d} N_{mi} \qquad (7.5)$$

and F_{am} is the allowable axial compressive stress given as a function slenderness ratio L_m/r_m

$$F_{am} = \begin{cases} \dfrac{\left[1 - \dfrac{(L_m/r_m)^2}{C_c^2}\right]F_y}{\dfrac{5}{3} + \dfrac{3(L_m/r_m)}{8C_c} - \dfrac{(L_m/r_m)^3}{C_c^3}} & \text{for } L_m/r_m \leq C_c \\ \dfrac{12\pi E}{23(L_m/r_m)^3} & \text{for } L_m/r_m \rangle C_c \end{cases} \qquad (7.6)$$

B. Stress and buckling constraints based on the AISC LRFD specifications

Similar to AISC ASD specifications, the stress constraints according to the AISC LRFD specifications can be expressed as

$$-F_{cm} \leq \sigma_{mk} \leq 0.9F_y, \quad m = 1,2\ldots,N_m \qquad (7.7)$$

where F_{cm} is the critical compressive strength given by

$$F_{cm} = \begin{cases} \left(0.658^{\lambda_c^2}\right)F_y & \text{for } \lambda_c = \frac{L_m}{r_m \pi}\sqrt{\frac{F_y}{E}} \leq 1.5 \\ \left(\frac{0.877}{\lambda_c^2}\right)F_y & \text{for } \lambda_c = \frac{L_m}{r_m \pi}\sqrt{\frac{F_y}{E}} > 1.5 \end{cases} \quad (7.8)$$

Considering only the active constraints the Lagrangian function for displacement constraints is defined as

$$L(\lambda_{jk}, A_i) = \sum_{i=1}^{N_d} \rho_i A_i \sum_{m=1}^{N_{mj}} L_{im} + \sum_{j=1}^{N_{ac}} \sum_{k=1}^{L} \lambda_{jk}(u_{jk} - r_j) \quad (7.9)$$

In order to obtain the optimality condition, we differentiate the Lagrangian function with respect to design variable A_i

$$\rho_i L_i + \sum_{j=1}^{N_{ac}} \sum_{k=1}^{L} \lambda_{jk} \frac{\partial u_{jk}}{\partial A_i} = 0 \quad (7.10)$$

Employing the principle of virtual load, the gradient of the jth displacement degree of freedom under the kth loading condition with respect to design variables A_i can be expressed as

$$g_{ijk} = \frac{\partial u_{jk}}{\partial A_i} = \frac{1}{A_i} \sum_{m=1}^{N_{mj}} -\mathbf{v}_{imj}^T \mathbf{k}_{im} \mathbf{u}_{imk} \quad (7.11)$$

where $\mathbf{v}_{imj}$ is the 6×1 vector of virtual displacements for member m belonging to group i due to the application of a unit load in the direction of the jth degree of freedom (three displacements for each node of the member), $\mathbf{k}_{im}$ is the 6×6 stiffness matrix of member m belonging to group i, and $\mathbf{u}_{imk}$ is the 6×1 displacement vector of member m belonging to group i due to the loading condition k.

After substituting the displacement constraint gradient, Eq. (7.11), into Eq. (7.10), rearranging terms and considering only the jth active displacement constraint under the kth loading condition, we find the Lagrange multipliers (Khot and Berke, 1984):

$$\lambda_{jk} = \frac{W}{u_{jk}} \tag{7.12}$$

Substituting Eqs. (7.11) and (7.12) into Eq. (7.10) and rearranging the terms, we obtain the following recurrence equation:

$$(A_i)_{p+1} = (A_i)_p \left\{ \sum_{c=1}^{N_{ac}} \left[(\lambda_{jk})_c \left(\frac{\sum_{m=1}^{N_{mj}} v_{imj}^T k_{im} u_{imk}}{\rho_i A_i \sum_{m=1}^{N_{mj}} L_{im}} \right) \right]_c \right\}_p^\varsigma , i = 1,2,....N_d \tag{7.13}$$

In this work, a value of 0.5 is initially used for ς in each iteration. If the weight of the structure is increased at the end of the iteration, a new value is used for ς equal to one half of the previous value. If the weight is decreased, the same $\varsigma = 0.5$ is used in the following iteration. A lower limit of 0.01 is used for ς. When ς is reduced to this value, in the next iteration ς is again initialized to 0.5. The idea behind a variable ς coefficient is to dampen the convergence oscillations.

For stress constraints, the cross-sectional areas are updated from the stress ratio relationship in the following form:

1. **For the case of AISC ASD specifications**

$$\left(A_m\right)_{p+1} = \left(A_m\right)_p \left[\frac{\sigma_{mk}}{F_{am} \text{ or } 0.6F_y}\right] \quad (7.14)$$

2. **For the case of AISC LRFD specifications**

$$\left(A_m\right)_{p+1} = \left(A_m\right)_p \left[\frac{\sigma_{mk}}{F_{am} \text{ or } 0.9F_y}\right] \quad (7.15)$$

In the multi-constraint optimization process, we first determine whether the most critical constraint (the most violated constraint) is a displacement constraint or a stress constraint. If a displacement constraint is a critical constraint, each member is first resized according to Eq. (7.13). Then, the maximum ratio of calculated displacement to the allowable displacement for various constraint degrees of freedom (SFD) is found. If this ratio is not equal to one within a given tolerance, then the design variables are scaled by this ratio. After the resizing and scaling, those members whose stress constraints are still violated are scaled again according to Eq. (7.14) or Eq. (7.15).

If a stress constraint is the critical constraint, all members are scaled according to Eq. (7.14) or Eq. (7.15). In the case of high-rise and super high-rise building structures the maximum lateral displacement (drift) is often the critical constraint. Thus, a resizing of the members according to Eq. (7.13) is performed in each iteration followed by the two aforementioned scaling operations.

For large structures with slender members, the allowable compressive stress (the nominal compressive strength in the case of design on the basis of the LRFD specifications) is much smaller than the allowable tensile stress (the nominal tensile strength in the case of design on the basis of the LRFD specifications). Furthermore, the constraints for compressive strength are nonlinear functions of the radii of gyration and implicit functions of design variables (cross-sectional areas). Scaling the members under compression may create a convergence problem. When the number of members whose compression constraints are violated is large, an "overshooting" problem may deteriorate the convergence performance of the algorithm. That is, after scaling of the members, they become substantially stiffer, resulting in a lateral drift substantially less than the allowable value. This can create a loop with a very slow convergence before we continue to calculate the Lagrange multipliers corresponding to active displacement constraints. To circumvent this problem, the scaling factor for the most violated displacement constraint (SFD) in the iteration is adjusted by a factor of γ whose value depends on the number of members whose compressive stress constraints are violated (N_c). The reason for introducing this parameter is to reduce the number of required structural analyses, which is of paramount importance for the efficiency of the algorithm for optimization of large structures. We found that by using $\gamma = \text{SFD}(0.9)(1 - N_c / N_m)$, the required number of structural analyses is reduced substantially.

There are seven loops in our multi-constraint optimality criteria approach for discrete optimization of a large structure, as shown schematically in Figure 7.1. The first loop is the main loop where the number of optimization iterations is controlled. In the second loop, the maximum scaling factor is obtained from the initial values for the cross sections. In the third loop, the design variables are estimated by satisfying the displacement constraints only, assuming continuous variables, and a new scaling factor is computed.

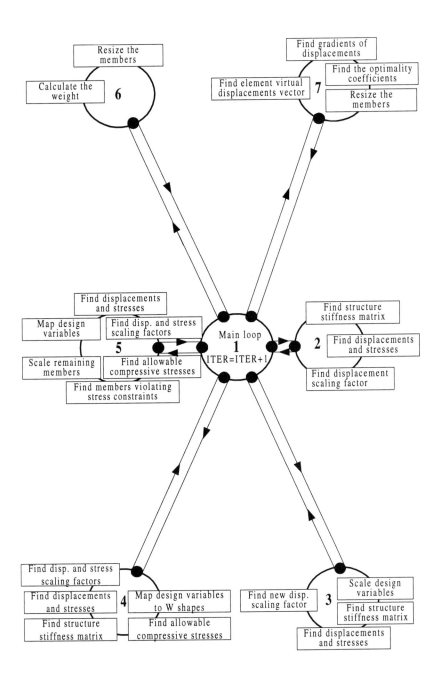

Figure 7.1: Multi-constraint optimality criteria discrete optimization

In the fourth loop, the continuous design variables are mapped to the available discrete standard cross sections from the AISC W sections database following a strategy to be discussed in the next section. The maximum displacement scaling factor (SFD) and scaling factors for those members whose compressive stress constraints are violated (SFSs) are obtained. In the fifth loop, all the members whose compressive stress constraints are violated are scaled by SFSs and the remaining members are scaled by γ. Then, all the members are mapped again to the discrete standard cross sections from the AISC W sections database. This process is repeated in loop 5 until all compression members satisfy the stress and buckling constraints.

The sixth loop is to calculate the weight of the structure and resize the members according to Eq. (7.13). If the weight of the structure in the current iteration is less than the weight in the previous iteration or the step size parameter ς is less than 0.01, the optimality coefficients are calculated and the members are resized in loop 7. Otherwise, the resizing procedure in loop 6 is repeated using a value of $\varsigma = \varsigma/2$.

7.3 MAPPING TO STANDARD CROSS SECTIONS

In the practical design of steel structures, only a finite number of shapes are available, such as those given in the AISC manuals (AISC, 1989&1994). We create a database containing the properties of the commonly used wide flange (W) shapes from these manuals in ascending order of the cross-sectional areas. In loops 4 and 5 of the discrete optimization algorithm (Figure 7.1), a W shape is selected for each group of members. For a large structure with a few thousand members (and a couple of hundred different types of design variables), the number of searches to select the right W shapes from the database can reach tens of thousands. Thus, an efficient strategy needs to be devised to map the computed cross-sectional areas to the existing W shapes from the database.

In this work, an integer search method is developed to map the computed cross-sectional areas (A_i) to the areas of actual W shapes in the database (A_i'). The integer search method is similar to the interval halving method (Adeli and Al-Rijleh, 1987). In the interval halving method, logical (if-then) operations are used to find the search path. In the integer search method, we use division operations to find the search path. The CPU time required for a division operation is a fraction of the CPU time required for a logical operation. Hence, the integer search method developed in this research is more efficient than the interval halving method.

Let us consider a 3-layer tree network with eight cross-sectional areas as target output as shown in Figure 7.2. Each node in this network has an upper and a lower branch. The kth node in the jth layer is assigned an area $A_{k,j}$ equal to the largest cross-sectional area of all the "child" nodes in the upper branch.

In the interval halving method, the search is performed as follows: If the computed cross-sectional area is less than or equal to $A_{1,0}$, the search path follows the upper branch; otherwise, it follows the lower branch. If the computed cross-sectional area is less than $A_{1,0}$, then the next step is to find out whether the computed cross-sectional area is less than or equal to $A_{1,1}$. If it is less than $A_{1,1}$, the search path follows the upper branch; otherwise, it follows the lower branch. If the computed cross-sectional area is less then or equal to $A_{2,2}$, the target output is $A_{3,3}$; otherwise, it is $A_{4,3}$.

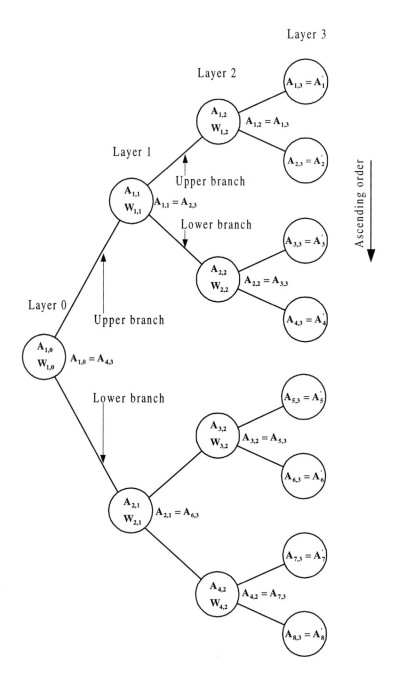

Figure 7.2: A three-layer tree network

In our integer search method, we assign a weight $W_{k,j}$ to the kth node in the jth layer of the tree network (layers are numbered from left to right, and nodes in each layer are numbered from top to bottom). These weights are stored in a two-dimensional array as shown in Table 7.1. The search is performed by finding the address (row number) of the weight in the successive layer that can lead to the right output. The address of the weight in the next layer is determined by the integer value resulting from division of the computed cross-sectional area by the weight of the current node. For instance, the address of the weight of the node in layer one on the solution path is determined by dividing the computed cross-sectional area by $W_{1,0}$. The integer value of this division plus one equals the address of the weight in layer one. If, for instance, the result of this division plus one is one, then the corresponding weight $W_{1,1}$ is stored in the first row and the second column of the array W. The next step is to find the address of the weight of the node on the search path in layer 2. This is obtained by dividing the computed cross-sectional area by $W_{1,1}$. If, for instance, the result of this division plus one is two, then the corresponding weight in layer two is $W_{2,2}$ and is stored in the second row and third column of the array W. If the result of the division is, say, 3, then the weight $W_{2,2}$ is stored in the third row and the third column of the array W. Finally, the computed cross-sectional area is divided by $W_{2,2}$. The result of this division determines the address of the target output. If the result of this division plus one is three, then the output is A'_3. If the result of the division plus one is four, then the output is A'_4. The problem now is how to determine the weight of each node and how to store them in a two-dimensional array that can lead to the right target output.

Table 7.1: Storage scheme for the weights and outputs

→ Column number

↓ Row number

		1	2	3	4
		Weight ($W_{k,j}$)			Output
		Layer 0	Layer 1	Layer 2	Layer 3
	1	$W_{1,0}$	$W_{1,1}$	$W_{1,2}$	$A_{1,3}$
	2		$W_{2,1}$	$W_{2,2}$	$A_{2,3}$
	3			$W_{3,2}$	$A_{3,3}$
	4			$W_{4,2}$	$A_{4,3}$
	5				$A_{5,3}$
	6				$A_{6,3}$
	7				$A_{7,3}$
	8				$A_{8,3}$

In Figure 7.2 the cross-sectional area of the first node of each layer j is assigned as the weight of that node, that is, $W_{1,j} = A_{1,j}$. If the computed cross-sectional area is less than $W_{1,j}$, then the search path follows the upper branch. Otherwise, the search path follows the lower branch. Let us consider the weight of the 1st node in the 1st layer, $W_{1,1}$. If the computed cross-sectional area (A_i) is less than $W_{1,1}$, then the search path follows the upper branch. The address of the weight, $W_{1,2}$, in the next layer is $\left(1 + \left\lfloor \dfrac{A_i}{W_{1,j}} \right\rfloor\right) = 1$, where the notation $\lfloor \ \rfloor$ indicates the integer of the real number inside the semi-brackets. If the computed cross-sectional area (A_i) is larger than $W_{1,1}$, then the search path follows the lower branch. The address of the weight $W_{2,2}$ in the next layer is $\left(1+\dfrac{A_i}{W_{1,j}}\right) \geq 2$. Thus, the number of addresses for the weight of $W_{2,2}$ may be more than one. This number depends on the largest cross-sectional area of all the "child" nodes in the lower branch, that is $A_{4,3}$. Let us assume $\left(1+\dfrac{A_{4,3}}{W_{1,j}}\right) = 3$. Then, $W_{2,2}$ is stored in both rows 2 and 3 and column 3 of array W as shown in Table 7.1. Thus, 3 is the maximum address number of $W_{2,2}$. The maximum address number for weight $W_{k,j}$ is called $L_{k,j}$.

Now, let us find the weights of other nodes. For instance, let us find the weight of the 2nd node of layer 1, $W_{2,1}$. From Figure 7.2, the path from W_{21} follows the upper branch if the computed cross-sectional area is greater than $A_{4,3}$ and less than or equal to $A_{6,3}$. Therefore, these two areas are important in determining the weight $W_{2,1}$. When the computed cross-sectional area is slightly greater than $A_{4,3}$ and is divided by $W_{2,1}$, the result of this division plus one has to be equal to the address of $W_{3,2}$. When the computed cross-sectional area is slightly greater than $A_{6,3}$ and is divided by $W_{2,1}$, the result of this division plus one has to be equal to the address of $W_{4,2}$. In order to satisfy these two conditions, the weight $W_{2,1}$ is determined in three steps as follows: First, $A_{4,3}$ is divided by the maximum address number of $W_{2,2}(L_{2,2})$, which is 3 in our example. The result of this division is called s. The weight of $W_{2,1}$ has to be less than s. Second, we compute the integer

$I = 1 + \left\lfloor \dfrac{A_{6,3}}{S} \right\rfloor$. Finally, the weight is computed as $W_{2,1} = A_{6,3}/I$. The maximum address number of the weight $W_{3,2}$ in the next layer is $L_{3,2} = \left(1 + \left\lfloor \dfrac{A_{6,3}}{W_{2,1}} \right\rfloor \right)$.

Now, we generalize the concepts presented in the previous paragraphs. Let us assume 2^n W shapes are available in the database where n is an integer. A tree network with n layer is created for the W shapes as shown in Figure 7.3. Each node is assigned an area from the W shape database according to the following equations:

$$A_{k,j} = A_i^{'} \quad i = 2^{n-j-1} + (k-1)*2^{n-j}, \quad k = 1,2^j, \ j = 1,......n \quad (7.16)$$

The next step is to assign a weight $W_{k,j}$ to the kth node in the jth layer of the tree network and find the maximum address number of the weights in the next layer $(L_{k,j+1})$. Each weight is stored in an array as shown in Table 7.1. For instance, if $L_{2,2}$ is equal to 3, then $W_{2,2}$ is stored in rows 2 and 3 and column 3 of array W.

The weight $(W_{k,j})$ and the maximum address numbers of the weight in the successive layers $(L_{k,j+1})$ are determined in three steps. The first step is to assign the weights of the first node of layers 1 to n-1 and find the maximum address number of the weights in the next layer as follows:

$$W_{1,j} = A_{1,j} \qquad j = 1,......,n-1 \qquad (7.17)$$

$$L_{1,j+1} = 1 \qquad j = 1,......,n-1 \qquad (7.18)$$

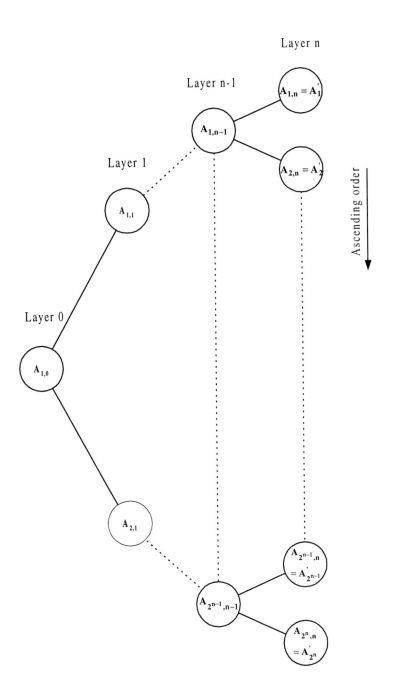

Figure 7.3: A tree network for mapping to standard sections

$$L_{2,j+1} = \left\lfloor \frac{A_{4,j+2}}{W_{1,j}} \right\rfloor + 1 \qquad j=1,......,n-1 \qquad (7.19)$$

The second step is to compute the weights of the nodes in layers 1 to n−2 and the maximum address numbers of the weights in the successive layers 2 to n−1. The weights are determined by the following rules:

$$s = \frac{A_{4k-4,j+2}}{L_{2k-2,j+1}} \qquad k=2,2^j \quad j=1,......,n-2 \qquad (7.20)$$

$$I = \left\lfloor \frac{A_{k,j}}{s} \right\rfloor \qquad k=2,2^j \quad j=1,......,n-2 \qquad (7.21)$$

$$W_{k,j} = \frac{A_{k,j}}{I+1} \qquad k=2,2^j \quad j=1,......,n-2 \qquad (7.22)$$

$$L_{2k-1,j+1} = \left\lfloor \frac{A_{4k-2,j+2}}{W_{k,j}} \right\rfloor + 1 \qquad k=2,2^j \quad j=1,......,n-2 \qquad (7.23)$$

$$L_{2k,j-1} = \left\lfloor \frac{A_{4k,j+2}}{W_{k,j}} \right\rfloor + 1 \qquad k=2,2^j \quad j=1,......,n-2 \qquad (7.24)$$

The third step is to compute the weights of the nodes in layer n−1 and the maximum address numbers of the outputs in layer n.

$$s = \frac{A_{2k-2,j+1}}{L_{2k-2,j+1}} \qquad k=2,2^j \quad j=1,......,n-1 \qquad (7.25)$$

$$I = \left\lfloor \frac{A_{k,j}}{s} \right\rfloor \qquad k=2,2^j \quad j=1,......,n-1 \qquad (7.26)$$

$$W_{k,j} = \frac{A_{k,j}}{I+1} \qquad k = 2, 2^j \quad j = 1, \ldots, n-1 \qquad (7.27)$$

$$L_{2k-1,+1} = \left\lfloor \frac{A_{2k-1,j+1}}{W_{k,j}} \right\rfloor + 1 \qquad k = 2, 2^j \quad j = 1, \ldots, n-1 \qquad (7.28)$$

$$L_{2k,j+1} = \frac{A_{2k,j+1}}{W_{k,j}} + 1 \qquad k = 2, 2^j \quad j = 1, \ldots, n-1 \qquad (7.29)$$

Then, the mapping of the input layer to the output layer is done as follows:

$$I_0 = 0 \qquad (7.30)$$

$$I_j = \left\lfloor \frac{A_i}{W_{I_{j-1}+1, j-1}} \right\rfloor \qquad j = 1, \ldots, n \qquad (7.31)$$

$$A_i = A'_{I_n} \qquad (7.32)$$

To illustrate the application of the integer search method, consider a tree network with three layers (n = 3) as shown in Figure 7.4. In layer 0 we have the computed cross-sectional area as an input and layer 3 is the target output layer consisting of 8 (2^3) W shapes from the AISC manuals. The smallest cross-sectional area in this example is 13.5 in.^{2} (87.1 cm^2) and the maximum cross-sectional area is 49.10 in.^{2} (316.8 cm^2). Each node is assigned a cross-sectional area for a W shape according to Eq (7.16). The weights are computed from Eqs. (7.17) to (7.27). The assigned cross-sectional areas and the weights of each node are shown in Figure 7.4 and the storage scheme is shown in Table 7.2.

The mapping is performed by dividing the computed cross-sectional area to the corresponding weight of each layer.

158 Optimization of Large Steel Structures Using Standard Cross Sections

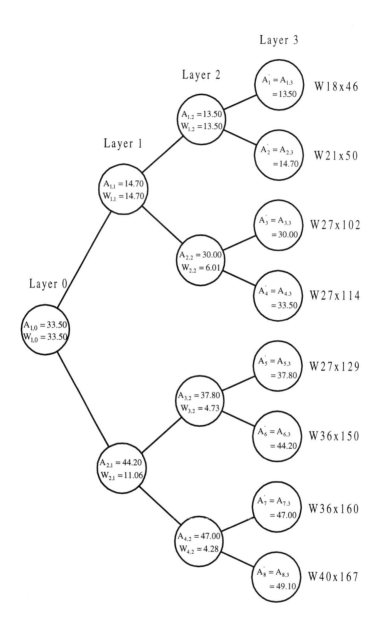

Figure 7.4: Example of a 3-layer network

Table 7.2: Storage scheme for weights and outputs

⟶ Column number

		1	2	3	4
		\multicolumn{3}{c}{Weight ($W_{k,j}$)}	Output
		Layer 0	Layer 1	Layer 2	Layer 3
Row number	1	38.80	14.70	13.50	13.50
	2	33.50	11.06	6.00	14.70
	3			6.00	30.00
	4			4.73	30.00
	5			4.28	30.00
	6				33.50
	7				33.50
	8				37.80
	9				44.20
	10				44.20
	11				47.00
	12				49.10

For instance, if the computed cross-sectional area is 40.0 in.^{2} (258.1cm^2) the address of the weight of the node in layer one is in row $\left(\left\lfloor\dfrac{40.0}{35.5}\right\rfloor+1\right)=2$ which indicates 11.06. The address of the weight of the node in layer 2 is in row $\left(\left\lfloor\dfrac{40.0}{11.06}\right\rfloor+1\right)=4$ which indicates 4.73. Finally the target output is in row $\left(\left\lfloor\dfrac{40.0}{4.73}\right\rfloor+1\right)=9$ which is $42.2\,\text{in.}^2\,\left(285.2\,\text{cm}^2\right)$ corresponding to W36 x 150.

7.4 APPLICATION

We present five examples in this section.

Example 1 72-member space truss

The 72 members of the 4-story space truss are divided into 16 design variables shown in Figure 7.5. This example has been solved by Adeli and Kamal (1986) and others and is presented for the sake of comparison only. This structure is subjected to five different loading cases. In the first case, the loading on the structure consists of two 22.24 kN (5 kip) forces in the x, y, and a -22.24 kN (-5 kip) force in the z direction at node 17. The second loading on the structure consists of two -22.24 kN (-5 kip) forces in the x, z and a 22.24 kN (5 kip) force in the y direction at node 18. The third loading case consists of three -22.24 kN (-5 kip) force in the x, y, and z directions at node 19. The fourth loading case consists of four -22.24 kN (-5 kip) forces in the z direction at nodes 17, 18, 19, and 20. Finally, the fifth loading case consists of two -22.24 kN (-5 kip) forces in the y and z directions and a 22.24 kN (5 kip) forces in the x direction at node 19. The member stresses are limited to 172.369×10^3 kPa (25 ksi) in both tension and compression. The maximum displacement in the x, y, z directions at the nodes on the top of the structure is limited to 0.63 cm (0.25 in). Modulus of elasticity of the material is 6.894×10^7 kPa (10000 ksi) and the unit weight of the material is 0.027 N/cm^3 (0.1 lb/in^3). The lower and upper

161

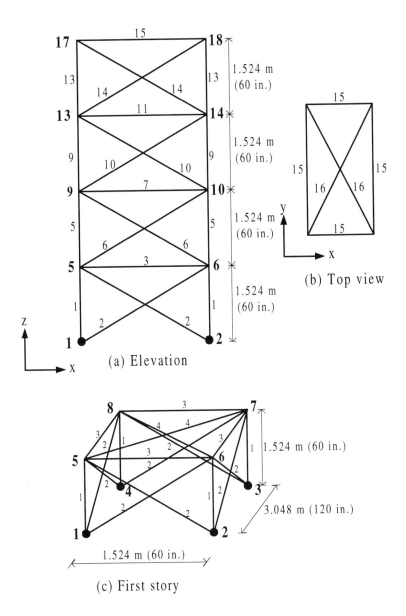

Figure 7.5: 72-member space truss with linking variables (numbers on the members refer to design variables and boldface numbers are node numbers)

bounds of the cross-sectional areas are 0.625 cm^2 and 500 cm^2. A minimum weight of 1690.2 N (380.01 lbs) is found in this work after three iterations. Venkayya (1971) reported a minimum weight of 1695.6 N (381.2 lbs) after 11 iterations. Adeli and Kamal (1986) obtained a minimum weight of 1687.2 N (379.31 lbs) after three iterations.

Example 2   52-story high-rise building

The 848-member space truss structure shown in Figures 7.6 to 7.8 is created to model the exterior envelope structure of a 52-story high-rise building (mega structure). Fifty-four design variables are used to represent the cross-sectional areas of fifty-four groups of members in this structure, as identified in Figure 7.7. The groups are organized as follows: Each four floors are divided into four groups: a group of outer column members, a group of inner column members, a group of diagonal members and a group of horizontal members.

The loading on the structure consists of horizontal loads acting on the exterior nodes of space structure at every four floors (14.63 m or 48 feet height). The horizontal loads in the y direction at each node on the sides AB and CD are obtained from the Uniform Building Code (UBC, 1994) wind loading using the equation $d_p = C_e C_q q_s I$, as defined in Section 5.7. The value of C_q for the inward face is 0.8 and for the leeward face is 0.5. Assuming a basic wind speed of 70 mph (113 km/h), the value of q_s is 0.6 kPa (12.6 psf) and the importance factor is assumed to be one. The values of C_e are taken from UBC (1994) assuming exposure C (generally open area). The lower and upper bounds of the cross-sectional areas of this example are 6 cm^2 and 1600 cm^2. This structure is designed according to the AISC ASD specifications (AISC, 1989). The material is

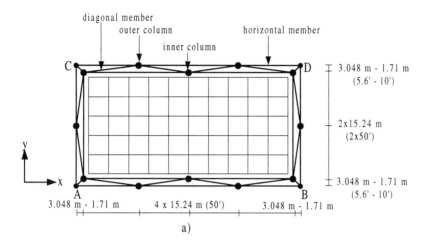

Figure 7.6: 52-story high-rise building structure

assumed to be steel with modulus of elasticity of 199.9×10^3 MPa (29,000 ksi) and the unit weight of material 0.077 N/cm^3 (0.284 lb/in^3). The displacement constraints are given as ± 78.1 cm (28.8 in.) in the y direction for the nodes on the top floor (equal to 0.004 H, where H is the height of the structure). Figure 7.9 shows the convergence histories for the cases of continuous variables optimization and discrete optimization using standard W shapes presented in this chapter. A minimum weight of 107.2 MN (24,108 kips) is found, assuming continuous variables. When W shapes are chosen on the basis of rounding up these values, and after performing additional optimization (scaling procedure) iteratively in order to satisfy the design constraints, a minimum weight of 123.3 MN (27,724 kips) is found. The discrete optimization algorithm presented in this chapter yielded a minimum weight of 112.0 MN (25,186 kips) directly after 8 iterations.

164 Optimization of Large Steel Structures Using Standard Cross Section

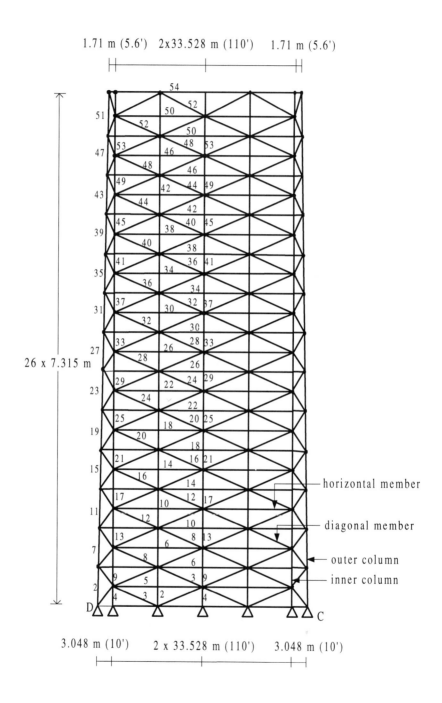

Figure 7.7: Front view of structure of Figure 7.6

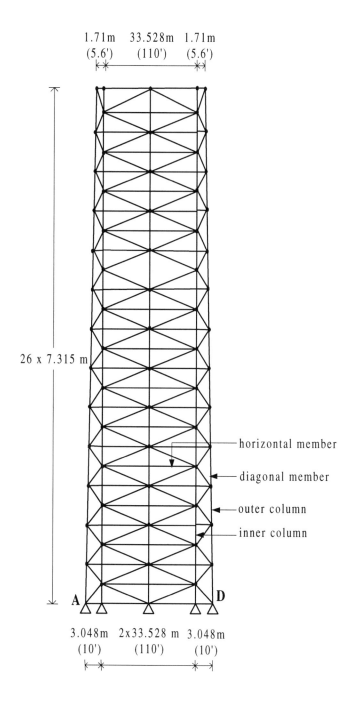

Figure 7.8: Side view of structure of Figure 7.6

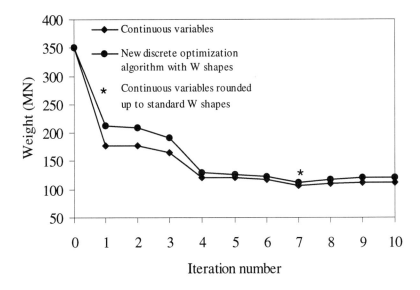

Figure 7.9: Convergence histories for the 52-story high-rise building structure

Example 3 35-story high-rise building structure

The 35-story space truss tower shown in Figures 7.10 and 7.11 consists of 1262 members and 324 nodes. Seventy-two design variables are used to represent the cross-sectional areas of seventy-two groups of members in this structure as shown in Figure 7.11. The loading on the structure consists of vertical loads (representing the dead and live loads) and horizontal loads (representing the wind loads). This structure consists of three different sections identified by 1 to 3 from the top to bottom. For the first section, the vertical dead (D) and live (L) loads are given as 7.1 kN (1.6 kips) and 6.2 kN (1.4 kips) at each node, respectively. For the second section, the vertical dead and live

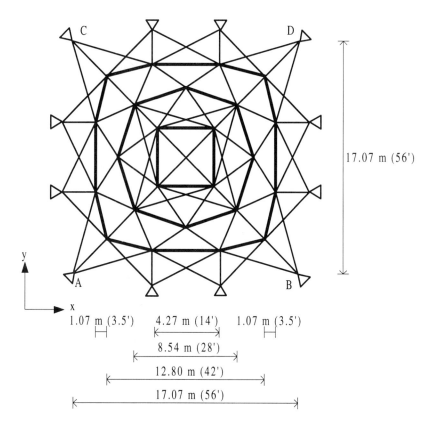

Figure 7.10: Plan of 35-story structure with linking variables

loads are given as 14.2 kN (3.2 kips) and 12.5 kN (2.8 kips) at each node, respectively. For the third section, the vertical dead and live loads are given as 21.3 kN (4.8 kips) and 18.6 kN (4.2 kips) at each node, respectively. Horizontal loads due to wind (W) in the y direction at each node on the sides AB and DC are calculated similar to example 2. This structure is designed using both AISC ASD and LRFD specifications. For the ASD specifications, only one case of loading is considered, that is, D + L + W. For the LRFD specifications, three cases of loading are considered: 1.4D, 1.2D + 1.6L + 0.5L_r and 1.2D + 1.3W +

168 Optimization of Large Steel Structures Using Standard Cross Section

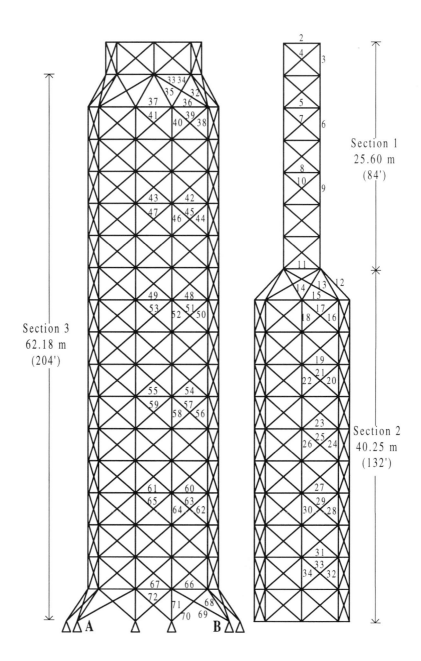

Figure 7.11 : Side view of the structure of Figure 7.10 (section 2 goes on the top of section 3)

0.5L + 0.5L$_r$. (L$_r$ is the roof live load). The lower and upper bounds of the cross-sectional areas in this example are 6 cm^2 and 6000 cm^2. The displacement constraints are given as ± 50.8 cm (20.0 in.) in the x and y directions for nodes on the top level (equal to 0.004H). For the LRFD specifications, the allowable horizontal displacement (drift) is increased by 30%. Figure 7.12 shows the convergence history for two minimum weight designs based on the ASD and LRFD specifications. A minimum weight of 4.22 MN (948.8 kips) is found after 8 iterations for the ASD specifications. A minimum weight of 3.94 MN (886.1 kips) is obtained after 6 iterations for the LRFD specifications.

Example 4 81-story super high-rise building structure

The 81-story super high-rise building structure shown in Figure 7.13 consists of 1160 nodes. This structure consists of two different sections as shown in Figures 7.13a to 7.13 d. We have used two slightly different models of the space structure in order to study the effect of the transition zone between the two sections in the overall efficiency of the structure. The model 4a as shown in Figure 7.13c has a sudden transition zone versus the model 4b as shown in Figure 7.13d which has a gradual transition zone. The latter model has 160 additional inclined members going from the 41st floor to the 48th floor. The model 4a has 5700 members grouped into 227 design variables. The model 4b has 5860 members grouped into 231 design variables. The groups are organized as follows: for section one, each two floors are divided into seven groups, i.e., a group of outer column members, a group of inner column members, a group of outer vertical diagonal members, a group of inner vertical diagonal members, a group of outer horizontal members, a group of inner horizontal members and horizontal diagonal

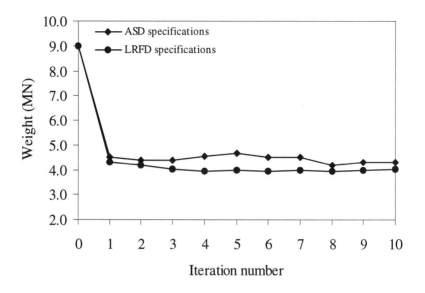

Figure 7:12: Convergence histories of the 35-story structure

members. For section two, each two floors are divided into four groups, that is, column members, vertical diagonal members, horizontal members and horizontal diagonal members.

The loading on section one consists of dead load (D) of 68.9 kN (15.5 kips) at each inner node and 10.14 kN (2.28 kips) at each outer node, live load (L) of 61.4 kN (13.80 kips) at each inner node and 12.54 kN (2.82 kips) at each outer node and roof live load (L_r) of 61.4 kN (13.80 kips) at each inner node and 12.54 kN (2.82 kips) at each outer node of the top level of the structure. The loading on section two consists of vertical dead loads of 58.7 kN (13.2 kips) at each node, live loads of 51.2 kN (11.5 kips) at each node and roof life loads of 51.2 kN (11.5 kips) at each node of the top level of the structure. This structure

171

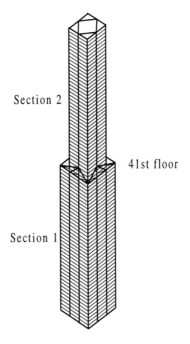

a: Perpective view of 81-story super high-rise building structure

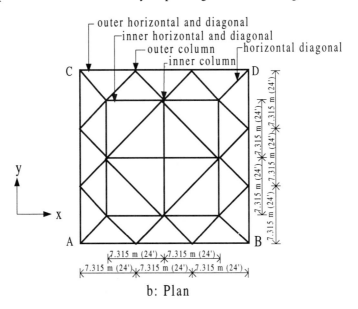

b: Plan

Figure 7.13: 81-story high-rise building structure

172 Optimization of Large Steel Structures Using Standard Cross Section

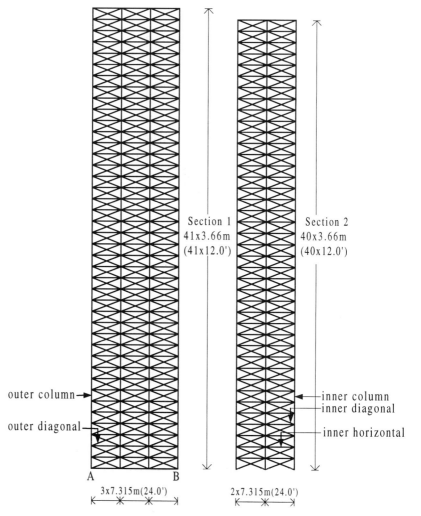

c: Elevation for example 4a (5700 members and 227 design variables

Figure 7.13-continued

173

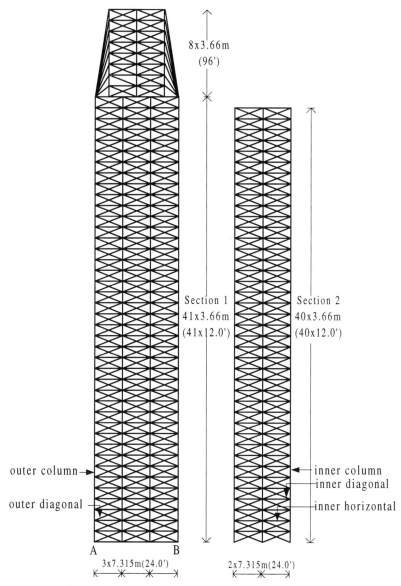

d: Elevation for example 4b (5860 members and 231 design variables)

Figure 7.13-continued

is subjected to horizontal wind loads in the y direction acting on each node on the sides AB and DC similar to example 3.

The structure is subjected to both AISC ASD and LRFD specifications; material properties are the same as the previous example. For the ASD specifications, only one case of loading is considered, that is, D + L + W. For the LRFD specifications, three cases of loadings are considered: 1.4D, 1.2D + 1.6L + 0.5L$_r$ and 1.2D + 1.3W + 0.5L + 0.5 L$_r$. The lower and upper bounds for the cross-sectional areas of this example are 6.0 cm^2 and 3000 cm^2. The displacement constraints are given as ± 118.36 cm (46.6 in.) in the y direction for the nodes on the top level (equal to 0.004H) for the ASD specifications and 153.8 cm (60.6 in.) for the LRFD specifications (equal to 0.0052H).

Figure 7.14 shows the convergence history for two different models of the structure with 227 and 231 design variables. For the ASD specifications, a minimum weight of 50.72 MN (11,430 kips) is found for model 4a (with 227 design variables) and a minimum weight of 43.53 MN (9,787 kips) is obtained for model 4b (with 231 design variables). For LRFD specifications, a minimum weight of 48.23 MN (10,842 kips) is found for model 4a and a minimum weight of 40.96 MN (9,208 kips) is obtained for model 4b.

Example 5 41-story high-rise building structure

This structure consists of the first 41 floors of example 4. It has 840 nodes and 4100 members grouped into 147 design variables. The loading on the structure and the material properties are the same as those of section one of example 4. The lower and upper bounds of the cross-sectional areas in this example are 6 cm^2 and 1500 cm^2. The structure is designed according to the AISC ASD and LRFD specifications. The

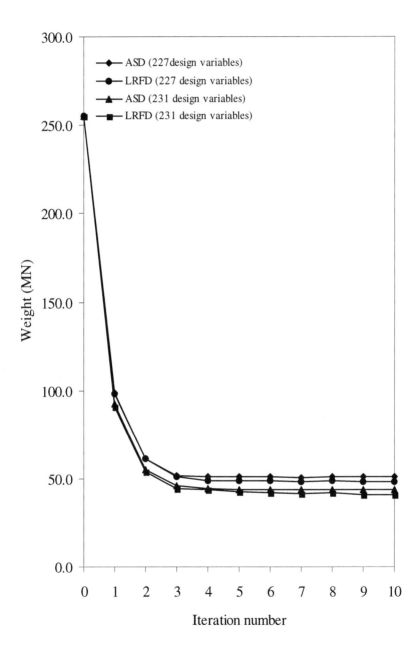

Figure 7.14: Convergence histories for the 81-story high-rise building structure

displacement constraints are given as ± 59.94 cm (23.6 in.) in the y direction for nodes on the top level (equal to 0.004H). For LRFD specifications, the horizontal displacement constraint is increased by 30%. Figure 7.15 shows the convergence history for two cases of design based on ASD and LRFD specifications. A minimum weight of 10.86 MN (2,442 kips) is found after 9 iterations for the ASD specifications. A minimum weight of 9.87 MN (2221 kips) is obtained after 9 iterations for LRFD specifications

7.5 FINAL COMMENTS

The multi-constraint discrete optimization algorithm presented in this chapter has been implemented in FORTRAN on a Cray YMP8/864 supercomputer. The code has been fully vectorized using the techniques described in Soegiarso and Adeli (1994). The efficiently vectorized code performs 16 to 19 times faster than the nonvectorized code.

The optimization convergence histories for various large structures presented in Figures 7.9, 7.12, 7.14 and 7.15 demonstrate consistently good performance and fast convergence. We also compared the integer search method developed in this research with the interval-halving approach (Adeli and Al-Rijleh, 1987). We found the former to be about 30 times faster than the latter for large structure. This is primarily due to the fact that the main operation in our integer search method requires much less time than the logical operation required in the interval-halving approach.

This research also sheds some light on the comparison of AISC ASD and LRFD specifications. Examples 3 to 5 are subjected to wind loading in addition to vertical dead and live loads. A maximum lateral drift of 0.004H was used for designs

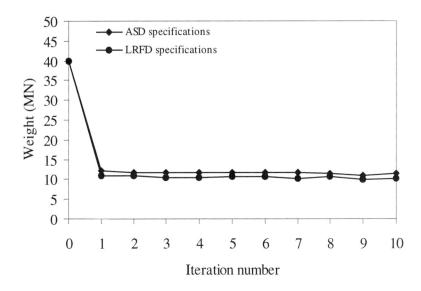

Figure 7.15: Convergence histories for the 41-story high-rise building structures

on the basis of the AISC ASD specifications. Since the drift is primarily due to wind loading and the load factor for wind loading in the AISC LRFD specifications is 1.3 (AISC, 1986), we increase the maximum allowable drift for designs based on the AISC LRFD specifications by 30%. We note that in all examples designs based on the LRFD specifications result in weight savings compared with designs based on the ASD specifications in the range of 5% to 9% (Figures 7.12, 7.14 and 7.15).

We also studied the effect of the inclined members in the transition zone of a super high-rise structure when the floor plan is reduced substantially. In example 4a, this transition is sudden versus the gradual transition in example 4b. Despite the fact that the number of design variables in example 4b is increased to

231 from 227 for example 4a, its minimum weight is about 14% (15%) less than the minimum weight for example 4a based on the AISC ASD specifications (based on the AISC LRFD specifications).

CHAPTER 8
Parallel-Vector Algorithm for Optimization of Large Structures

8.1 INTRODUCTION

The goal of this chapter is to present efficient and robust parallel-vector algorithms for optimization of large space steel structures subjected to the actual design constraints of the AISC ASD specifications (AISC, 1989). Some of the AISC specifications are highly nonlinear and implicit functions of design variables, especially in the case of moment-resisting frames. This can create convergence and stability problems and increase CPU requirements significantly, especially for large structures. The types of structures considered are space axial load structures, moment-resisting frames, and moment-resisting frames with diagonal bracings.

Extending the work presented in the previous chapter, we first formulate an optimality criteria approach for optimization of steel structures subjected to stress, displacement, and buckling constraints of AISC ASD specifications. Two different scaling procedures are used, one suitable for axial-load structures, and the other suitable for space moment-resisting frames. Then, the parallel-vector structural optimization algorithm is presented. Next, the algorithm is applied to minimum weight design of three space axial-load structures, three space moment-resisting frames, and one space moment-resisting frame with diagonal bracings. Finally, the performance of the algorithm is evaluated by presenting speedup results for parallel processing and vectorization.

8.2. AN OPTIMALITY CRITERIA APPROACH

In this section, we present an optimality criteria (OC) approach for optimization of steel structures subject to stress, displacement and buckling constraints based on the AISC allowable stress design (ASD) specifications (AISC, 1989). The general structural optimization problem with design linking strategy can be stated as in Eq. (7.1).

The displacement and fabrication constraints are the same as Eq. (6.19) and Eq. (6.20), respectively. The stress constraints are obtained from the AISC ASD specifications. When the beam-column member is under compression, the stress constraints are

$$\begin{cases} \dfrac{f_{amk}}{F_{am}} + \dfrac{C_{mx} f_{bmkx}}{\left(1 - \dfrac{f_{am}}{F'_{ez}}\right) F_{bmx}} + \dfrac{C_{my} f_{bmky}}{\left(1 - \dfrac{f_{am}}{F'_{ey}}\right) F_{bmy}} \leq 1.0 & \text{for } \dfrac{f_{amk}}{F_{am}} \rangle 0.15 \\ \dfrac{f_{amk}}{0.6 F_y} + \dfrac{f_{bmkx}}{F_{bmx}} + \dfrac{f_{bmky}}{F_{bmy}} \leq 1.0 \end{cases} \quad (8.1)$$

$$\frac{f_{amk}}{F_{am}} + \frac{f_{bmkx}}{F_{bmx}} + \frac{f_{bmky}}{F_{bmy}} \le 1.0 \qquad \text{for } \frac{f_{amk}}{F_{am}} \le 0.15 \quad (8.2)$$

where f_{amk} is the computed compressive stress in member m due to the loading condition k, f_{bmk} is the computed bending stress at the point under consideration due to loading condition k, F_{bm} is the allowable bending stress in member m, and F_y is the yield stress of the material. For compression members in unbraced frames, the value of C_m is 0.85. N_m is the total number of members in the structure which is equal to

$$N_m = \sum_{i=1}^{N_d} N_{mi} \qquad (8.3)$$

We assume full lateral support is provided for horizontal members (beams). For columns and inclined members, we assume lateral support is provided at the ends of members only. The term F_e' is defined as

$$F_e' = \frac{12\pi^2 E}{23(KL_b / r_b)_m^2} \qquad (8.4)$$

where L_b is the unbraced length in the plane of bending, r_b is the radius of gyration in the plane of bending, E is the modulus of elasticity of steel, and K is the effective length factor.

The allowable bending stress is computed based on Chapter F of the AISC ASD specifications. The allowable compressive stress, F_{am}, is given as a function of the governing slenderness ration for member m, $(KL/r)_m$, as shown in Eq. (7.6). The coefficient C_c is defined as

$$C_c = \sqrt{\frac{2\pi^2 E}{F_y}} \tag{8.6}$$

The effective length factor, K for braced and unbraced frames is found by the following approximate equations from the European steel design code (Anonymous, 1978, and Dumonteil, 1992):

$$K = \frac{3G_A G_B + 1.4(G_A + G_B) + 0.64}{3G_A G_B + 2.0(G_A + G_B) + 1.28} \quad \text{(for braced frames)} \tag{8.7}$$

$$K = \sqrt{\frac{1.6G_A G_B + 4.0(G_A + G_B) + 7.50}{G_A + G_B + 7.50}} \quad \text{(for unbraced frames)} \tag{8.8}$$

where subscripts A and B refer to the two ends of the column under consideration. The restraint factor G is defined as

$$G = \frac{\sum (I_c / L_c)}{\sum (I_g / L_g)} \tag{8.9}$$

where I_c is the moment of inertia and L_c is the unsupported length of column section, I_g is the moment of inertia and L_g is the unsupported length of girder, and $\sum$ indicates a summation for all members connected to the joint under consideration and lying in the plane of buckling.

When the member is under tension, the stress constraint takes the following form:

$$\frac{f_{amk}}{0.60 F_t} + \frac{f_{bmkz}}{F_{bmz}} + \frac{f_{bmky}}{F_{bmy}} \leq 1.0 \tag{8.10}$$

where $F_t = 0.60 F_y$ is the allowable tensile stress.

Considering only the active constraints, the Lagrangian function for displacement constraints is defined as in Eq. (7.9). Employing the principle of virtual work, the gradient of the jth displacement degree of freedom under the kth loading condition with respect to design variable A_i can be expressed as

$$g_{ijk} = \frac{\partial u_{jk}}{\partial A_i} = \sum_{m=1}^{N_{mj}} - v_{imj}^T \frac{k_{im}}{\partial A_i} u_{imk} \qquad (8.11)$$

where v_{imj} is the virtual displacement vector of member m belonging to the group i due to the application of a unit load in the direction of the jth degree of freedom. For a space frame, v_{imj} is a 21x1 vector (three displacements and three rotations for each node of the member) and for a space truss, v_{imj} is a 6x1 vector (three displacements for each node). k_{im} is the 12x12 (6x6) stiffness matrix of member m belonging to group i for a space frame (space truss). u_{imk} is the 12x1 (6x1) displacement vector of member m belonging to group i due to the loading condition k for a space frame (space truss). For a space frame, k_{im} is composed of two contributions: one corresponding to axial forces (k_{ima}) and the other corresponding to bending moments (k_{imb}). Thus Eq. (8.11) can be written as

$$g_{ijk} = \frac{\partial u_{jk}}{\partial A_i} = \sum_{m=1}^{N_{mj}} - v_{imj}^T \frac{k_{ima} + k_{imb}}{\partial A_i} u_{imk} \qquad (8.12)$$

We use two different scaling procedures, one for space truss (axial load) structures and the other for space moment-resisting frames. For space truss structures, the member stiffness matrix is a function of the cross-sectional area and we use a scaling procedure based on a combination of the maximum displacement scaling factor (SFD) and stress scaling factors for those members whose compressive stress constraints are violated (SFSs). Details of this scaling procedure are given in Chapter 7. For space frames, the member stiffness matrix is a function of not only the cross-sectional area but also the moments of inertia I_x, and I_y and the torsion constant J. Therefore, a different scaling procedure is devised.

In the case of high-rise building structures we know a priori that the lateral displacements are normally the critical displacement constraints. The lateral displacements are induced primarily by horizontal loads (wind or earthquake loads). In the moment-resisting unbraced frames the lateral displacements are controlled largely by bending actions of beams and columns, that is the moments of inertia of the cross sections. When a displacement constraint is violated, we first compute the maximum displacement scaling factor (SFD) as

$$SFD = \max \frac{u_{jk}}{r_j'} \text{ for } j = 1, \ldots, N \quad k = 1, \ldots, L \qquad (8.13)$$

Then, the new moments of inertia of the members are obtained from the following relationship:

$$I'_{mx} = SFD * I_{mx} \qquad (8.14)$$

where I'_{mx} and I_{mx} are the new and previous moments of inertia of member m, respectively.

Next, the computed moment of inertia of each member is mapped to the moment of inertia of a W shape in the AISC's database using the mapping strategy presented in Chapter 7. If the moments of inertia of the selected W shapes are all larger than the scaled moments of inertia (which is often the case), the structure becomes substantially stiffer resulting in a lateral drift substantially less than the allowable value. This can create a loop with a very slow convergence before we continue to calculate the Lagrange multipliers corresponding to the active displacement constraints. To circumvent this problem, we perform up to three scalings. The displacement scaling factor for the most violated displacement constraint (SFD) in the first scaling in each iteration is adjusted by a factor γ_1. In the second scaling, only 50% of the members from the top of the structure are scaled using SFD multiplied by a factor γ_2. After the second scaling, if the maximum lateral drift is still less than the allowable value, the scaling procedure is continued for the 25% of the members from the top of the structure. The scaling factor SFD is adjusted this time by a factor γ_3. We found the values $\gamma_1 = 0.90, \gamma_2 = 0.90$ and $\gamma_3 = 0.80$ to yield the best convergence results.

8.3 PARALLEL-VECTOR ALGORITHM

The impact of vectorization on structural optimization has been discussed in Chapter 6. Through the judicious use of the vectorization techniques, the speedup of the vectorized algorithm can reach the range 15 to 19. In this section, we present a parallel-vector multi-constraint discrete optimization algorithm for optimization of large steel structures.

In OC-based structural optimization algorithms, most of the computer processing time is spent in the assembling the structure stiffness matrix, computing gradients, and solving for nodal displacements. Most tasks in these steps are executed in a DO loop. Parallelization at a DO loop is called Microtasking (Section 5.2). The challenge is to develop algorithms where the outer DO loops can be partitioned and distributed to all processors evenly and the inner DO loops can be vectorized. In developing an efficient microtasked program it is necessary to understand how the values in the DO loop are stored. Basically, we have two kinds of storage. The first is global storage (shared memory) where the data can be used and modified in any subroutine and passed through the Common blocks argument. The second type is the local storage (private memory) where the data are used and modified in a subroutine only. In this case, the data cannot be passed to another subroutine. When the data stored in the shared memory are modified by different processors, a racing condition may be encountered resulting in erroneous results (Adeli and Kamal, 1993). To avoid undesirable results we employ the directives *Guard* and *Endguard* provided by the Cray FORTRAN compiler in a *guarded region*. In this region only one processor at a time can update the same location. The algorithm is as follows.

Step 1
Read in the input data such as the initial cross-section properties, tolerance for displacement constraints, and the step size parameters (sequential)

Step 2
Compute the member stiffness matrices
If the structure type = 1(space frame)
    {Parallel region-entry}
        For s =1 until P *(microtasking)*
            $m_1 = (s-1) * N_m / P + 1$
            $m_2 = s * N_m / P$
            For member m = m_1 until m_2 *(vectorization)*
                Calculate the transformation stiffness matrix
                $\mathbf{T}_{im}(12,12,m)$
                Calculate the member stiffness matrix
                $\mathbf{k}_{im\,l}(12,12,m)$ in local coordinates
                Calculate the member stiffness matrix
                $\mathbf{k}_{im}(12,12,m)$ in global coordinates

$$\mathbf{k}_{im} = \mathbf{T}_{im}^T \mathbf{k}_{iml} \mathbf{T}_{im}$$

Next member

Next s

Assemble the structure stiffness matrix

    {Guard region-entry}

        For member m =1 until N_m (*microtasking*)

            Assemble the member stiffness matrix into the structure stiffness matrix

$$\mathbf{K} = \sum_{m=1}^{N_m} \mathbf{k}_{im}(12,12,m)$$

        Next member

    {Guard region-end}

    {Parallel region-end}

Elseif the structure type=2 (*space truss*)

    {Parallel region-entry}

        For s=1 until P (*microtasking*)

        $m_1 = (s-1) * N_m / P + 1$

        $m_2 = s * N_m / P$

        For member m = m_1 until m_2 (*vectorization*)

            Calculate the member stiffness matrix

            $\mathbf{k}_{im}(6,6,m)$

        Next member

    Next s

Assemble the structure stiffness matrix

    {Guard region-entry}

        For member m = 1 until N_m (*microtasking*)

            Assemble the member stiffness matrix into structure stiffness matrix

$$\mathbf{K} = \sum_{m=}^{N_m} \mathbf{k}_{im}(6,6,m)$$

    Next member

    {Guard region-end}

{Parallel region-end}
Endif
Modify the structure stiffness matrix to include the boundary conditions
    {Parallel region-entry}
        For s=1 until P (*microtasking*)
$$j_1 = (s-1)*N/P + 1$$
$$j_2 = s*N/P$$
        For nodal degree of freedom j = j_1 until j_2
(*vectorization*)
            Modify the structure stiffness matrix to include the boundary conditions
        Next j
    Next s
{Parallel region-end}

Step 3
Assemble the loads
    For loading condition k = 1 until L
    {Parallel region-entry}
        For s=1 until P (*microtasking*)
$$j_1 = (s-1)*N/P + 1$$
$$j_2 = s*N/P$$
        For nodal degree of freedom j = j_1 until j_2
(*vectorization*)
$$\mathbf{F}_k = \sum_{j=1}^{N_m} f_{jk}$$
        Next j
    Next s
{Parallel region-end}
Next loading

Step 4
  *Factorize the structure stiffness matrix **K*** (as in Table 5.1 of Chapter 5)
  $$\mathbf{K} = \mathbf{L}\mathbf{L}^T$$

Step 5
  Forward substitution ($\mathbf{Lu}_k = \mathbf{F}_k$)
    For loading condition k = 1 until L
      For j = 1 until N
        $u_{jk} = f_{jk} / L_{jj}$
        For i = j+1, min(j+b,N) (*vectorization*)
          $u_{ik} = u_{ik} - L_{ij} * u_{jk}$
        Next i
      Next j
    Next loading

  Backward substitution ($\mathbf{Lu}_k = \mathbf{u}_k$)
    For loading condition k = 1 until L
      For j = N until 1, −1
        $u_{jk} = u_{jk} / L_{jj}$
        For i = j−1, min(j−b,1), −1 (*vectorization*)
          $u_{ik} = u_{ik} - L_{ji} * u_{jk}$
        Next i
      Next j
    Next loading

Step 6
If structure type = 1 (*space frame*)
Find the member forces
    For loading condition k = 1 until L
    {Parallel region-entry}
        For s=1 until P (*microtasking*)

$$m_1 = (s-1)*N_m/P+1$$
$$m_2 = s*N_m/P$$

For member m = m_1 until m_2 (*vectorization*)

Calculate the member displacements vector in local coordinates
$$\mathbf{u}_{imk\,l} = \mathbf{T}_{im}^T \mathbf{u}_{imk} \mathbf{T}_{im} \quad (unrolling)$$

Calculate the member forces
$$\mathbf{F}_{imk\,l} = \mathbf{k}_{im\,l} \mathbf{u}_{imk\,l} \quad (unrolling)$$

Add the member fixed-end moment
$$\mathbf{F}_{imk\,l} = \mathbf{F}_{imk\,l} + \mathbf{F}_{imk}$$

Next member

Next s

{Parallel region-end}

Next loading

If structure type = 2(*space truss*)

Find the member forces and stresses

For loading condition k = 1 until L

{Parallel region-entry}

For s=1 until P (*microtasking*)
$$m_1 = (s-1)*N_m/P+1$$
$$m_2 = s*N_m/P$$

For member m = m_1 until m_2 (*vectorization*)

Calculate the member elongation (δ_{mk})

Calculate the member stress
$$\sigma_{mk} = \delta_{mk} \frac{E}{L_m}$$

Calculate the member force
$$\mathbf{F}_{imk\,l} = \sigma_{imk} * A_{im}$$

Next member

Next s

{Parallel region-end}

Next loading

Step 7
If structure type = 1 (*space frame*)
Find the maximum displacement scaling factor
    For loading condition k = 1 until L
        For nodal degree of freedom j = 1 until N (*sequential*)
        Find the maximum displacement scaling factor $(SFD)_k$

$$(SFD)_k = \max \frac{u_{jk}}{r_j}$$

    Next j
    Next loading
Find the stress scaling factors for members
    For loading condition k = 1 until L
    {Parallel region-entry}
        For s=1 until P (*microtasking*)
        $m_1 = (s-1) * N_m / P + 1$
        $m_2 = s * N_m / P$
        For member m = m_1 until m_2 (*vectorization*)
            If $f_a \leq 0$ (*compression member*)

            Compute the value of the interaction equation and use it as the stress scaling factor

$$SF_{mk} = \max \begin{cases} \dfrac{f_{amk}}{F_{am}} + \dfrac{C_{mx}f_{bmkz}}{\left(1-\dfrac{f_{am}}{F'_{ex}}\right)F_{bmz}} + \dfrac{C_{my}f_{bmky}}{\left(1-\dfrac{f_{am}}{F'_{ey}}\right)F_{bmy}} \\ \dfrac{f_{amk}}{F_{am}} + \dfrac{f_{bmkz}}{F_{bmz}} + \dfrac{f_{bmky}}{F_{bmy}} \end{cases}$$

$$\text{for } \frac{f_{amk}}{F_{am}} \leq 0.15$$

$$SF_{mk} = \frac{f_{amk}}{0.60F_y} + \frac{f_{bmkz}}{F_{bmz}} + \frac{f_{bmky}}{F_{bmy}} \quad \text{for } \frac{f_{amk}}{F_{am}} \rangle 0.15$$

Elseif $f_a \rangle 0$ (*tension member*)

Compute the value of interaction equation and use it as the stress scaling factor

$$SF_{mk} = \frac{f_{amk}}{0.60F_t} + \frac{f_{bmkz}}{F_{bmz}} + \frac{f_{bmky}}{F_{bmy}}$$

    Endif
   Next member
  Next s
{Parallel region-entry}
Next loading

Find the maximum stress scaling factor
    For loading condition k = 1 until L
        For member m = 1 until N_m (*sequential*)
            Find the maximum stress scaling factor $(SFS)_k$

$$(SFS)_k = \max SF_{mk}$$

    Next member
  Next loading
If the structure type = 2 (*space truss*)

Find the maximum displacement scaling factor
    For loading condition k = 1 until L
        For nodal degree of freedom j = 1 until N (*sequential*)
            Find the maximum displacement scaling factor $(SFD)_k$

$$(SFD)_k = \max \frac{u_{jk}}{r_j}$$

            Next nodal degree of freedom
      Next loading

Find the stress scaling factors
      For loading condition k = 1 until L
         {Parallel region-entry}
         For s=1 until P (*microtasking*)
            $m_1 = (s-1) * N_m / P + 1$
            $m_2 = s * N_m / P$
            For member m = m_1 until m_2 (*vectorization*)
               If $f_{amk} \leq 0$ (*compression member*)
                  $$SF_{mk} = \frac{f_{amk}}{F_a}$$
               Elseif $f_a \rangle 0$ (*tension member*)
                  $$SF_{mk} = \frac{f_{amk}}{0.60F_y}$$
               Endif
            Next member
         Next s
         {Parallel region-end}
      Next loading

Find the maximum stress scaling factor
      For loading condition k = 1 until L
         For member m = 1 until N_m (*sequential*)
            Find the maximum stress scaling factor $(SFS)_k$
            $(SFS)_k = \max SF_{mk}$
         Next member
      Next loading

Step 8
If the structure type = 1 (*space frame*)

Scale the moments of inertia and select a W shape
   If $|SFD - 1.0| \rangle$ tol. and $N_{an} = 1$
     For loading condition k
     {Parallel region-entry}
        For s=1 until P (*microtasking*)
$$m_1 = (s-1)*N_d/P + 1$$
$$m_2 = s*N_d/P$$
           For member m = m_1 until m_2 (*vectorization*)
$$I_{mx} = I_{mx}*(SFD)_k*\gamma_1$$
           Next member
        Next s
   {Parallel region-end}
   {Parallel region-entry}
        For s=1 until P (*microtasking*)
$$m_1 = (s-1)*N_d/P + 1$$
$$m_2 = s*N_d/P$$
           For member m = m_1 until m_2 (*vectorization*)
              Select W shapes using the mapping strategy described in Chapter 7
           Next member
        Next s
   {Parallel region-end}
     $N_{an} = N_{an} + 1$
Endif
Go to Step 2
If $|SFD - 1.0| \rangle$ tol. and $N_{an} = 2$
   For loading condition k
   {Parallel region-entry}
        For s=1 until P (*microtasking*)
$$m_1 = N_d/2 + (s-1)*N_d/2P + 1$$
$$m_2 = N_d/2 + s*N_d/2P$$

For member m = m_1 until m_2 (*vectorization*)
   $$I_{mx} = I_{mx} * (SFD)_k * \gamma_2$$
   Next member
   Next s
{Parallel region-end}
{Parallel region-entry}
   For s=1 until P (*microtasking*)
   $$m_1 = N_d/2 + (s-1)*N_d/2P + 1$$
   $$m_2 = N_d/2 + s*N_d/2P$$
   For member m = m_1 until m_2 (*vectorization*)
   Select W shapes using the mapping strategy described in Chapter 7
   Next member
   Next s
{Parallel region-end}
   $$N_{an} = N_{an} + 1$$
Endif
Go to step 2
If $|SFD - 1.0| >$ tol. and $N_{an} = 3$
For loading condition k
{Parallel region-entry}
   For s =1 until P (*microtasking*)
   $$m_1 = 3*N_d/4 + (s-1)*N_d/4P + 1$$
   $$m_2 = 3*N_d/4 + s*N_d/4P$$
   For member m = m_1 until m_2 (*vectorization*)
   $$I_{mx} = I_{mx} * (SFD)_k * \gamma_3$$
   Next member
   Next s
{Parallel region-end}
{Parallel region-entry}
   For s=1 until P (*microtasking*)

$$m_1 = 3*N_d/4 + (s-1)*N_d/4P + 1$$
$$m_2 = 3*N_d/4 + s*N_d/4P$$

      For member m = m_1 until m_2 (*vectorization*)

        Select W shapes using the mapping strategy described in Chapter 7

      Next member

    Next s

{Parallel region-end}

    $N_{an} = N_{an} + 1$

Endif

Go to step 2

If $|SFD - 1.0| >$ tol. and $3 < N_{an} \le 5$

Find those members whose stress constraints are violated

{Parallel region-entry}

    For s=1 until P (*microtasking*)

      $i_1 = (s-1)*N_d/P + 1$

      $i_2 = s*N_d/P$

      For i = i_1 until i_2 (*sequential*)

        $N_c = 0$

        For member m = 1 until N_{mi}

          If $SF_{mk} > SF$

            $SF = SF_{mk}$

            $N_c = N_c + 1$

            im = i

          Endif

        Next member

        If $(N_c > 1)$ then

          $I_{imx} = I_{imx} * SF * 1.1$

        Endif

        Select W shapes using the mapping strategy described in Chapter 7

        Next i
      Next s
{Parallel region-end}
    $N_{an} = N_{an} + 1$

Endif

Go to step 2

If structure type = 2 *(space truss)*
SFD $\rangle$ SFS

If $|SFD - 1.0| \rangle$ tol. and $N_{an} = 0$

Scale the cross-sectional areas
    {Parallel region-entry}
      For s=1 until P *(microtasking)*
        $m_1 = (s-1) * N_m / P + 1$
        $m_2 = s * N_m / P$
        For member m = m_1 until m_2 *(vectorization)*
          $A_m = (SFD)_k * A_m$
      Next member
    Next s
{Parallel region-end}
  $N_{an} = N_{an} + 1$

Endif

Go to step 2

If $|SFD - 1.0| \langle$ tol. and $|SFS - 1.0| \rangle$ tol

Select a W shape
    {Parallel region-entry}
      For s=1 until P *(microtasking)*
        $m_1 = (s-1) * N_d / P + 1$
        $m_2 = s * N_d / P$
        For member m = m_1 until m_2 *(vectorization)*
          Select a W shape using the mapping strategy described in Chapter 7

                Next member
            Next s
        {Parallel region-end}
        $N_{an} = N_{an} + 1$
        Endif
        Go to step 2
If $|SFD - 1.0| >$ tol. and $N_{an} > 1$
Find those members whose stress constraints are violated
        {Parallel region-entry}
            For s=1 until P (*microtasking*)
                $i_1 = (s-1) * N_d / P + 1$
                $i_2 = s * N_d / P$
                For i = i_1 until i_2 (*sequential*)
                    For member m = 1 until N_{mi}
                        If $SF_{mk} > SF$
                            $SF = SF_{mk}$
                            $N_c(s) = N_c(s) + 1$
                        Endif
                    Next member
                    Scale those members whose stress constraints are violated
                    Scale the remaining members by the adjusted scaling factor described in Chapter 7
                    $A_i = SF * A_i$
                Next i
            Next s
        {Parallel region-end}
        For s = 1 until P (*sequential*)
            $N_c = N_c + N_c(s)$
        Next s
        $N_{an} = N_{an} + 1$

Endif
Go to step 2
If $\left(|SFD-1.0| \leq \text{tol. and } N_c = 0 \text{ or } N_{am} \rangle 5\right)$ go to step 9

Step 9
Calculate the weight of the structure
{Parallel region-entry}
    For s=1 until P (*microtasking*)
        $m_1 = (s-1) * N_m / P + 1$
        $m_2 = s * N_m / P$
        For member m = m_1 until m_2 (*vectorization*)
$$W_1(s) = \sum_{m=m_1}^{m_2} A_m * L_m * \rho_m$$
        Next member
    Next s
{Parallel region-end}
For s = 1 until P (*sequential*)
    $W = W + W_1(s)$
Next s

Step 10
Check the weight of the structure
    If $\left(W^{new} \rangle W^{old} \text{ and } \zeta \rangle 0.1\right)$ $\zeta = 0.5 * \zeta$ and go to step 14
ITER = ITER +1

Step 11
Find the active displacement constraints and their corresponding Lagrange multipliers λ_{jk}

Step 12
Calculate the virtual nodal displacements
    For active displacements a = 1 until N_{ac}

Find the virtual nodal displacements due to a unit load in the direction of the jth active displacement constraint (as in step 5)

$$\mathbf{K}\mathbf{v}_{ja} = \mathbf{F}_{ja}$$

Next active constraint

For j = 1 until N_{ac}

{Parallel region-entry}

For s=1 until P (*microtasking*)

$m_1 = (s-1) * N_m / 4P + 1$

$m_2 = s * N_m / 4P$

For member m = m_1 until m_2 (*vectorization*)

Find the member virtual nodal displacements $\mathbf{v}_{imj}$

Next member

Next s

{Parallel region-end}

Next j

Step 13

Calculate displacement gradients

{Parallel region-entry}

For s=1 until P (*microtasking*)

$m_1 = (s-1) * N_m / P + 1$

$m_2 = s * N_m / P$

For member m = m_1 until m_2

Calculate member stiffness matrix $\mathbf{k}(12,12,m)$

(*space frame*)

or $\mathbf{k}(6,6,m)$ (*space truss*)

Next member

Next s

{Parallel region-end}

For loading condition k

{Parallel region-entry}

For i = 1 until N_d (*microtasking*)
    For member m = 1 until N_{mi} (*vectorization*)
        Calculate member displacements vector $\mathbf{u}_{imk}$
    Next member
Next i
{Parallel region-end}
For loading condition k
    For j = 1, N_{ac}
        {Parallel region-entry}
        For i = 1 until N_d (*microtasking*)
            For member m = 1 until N_{mi} (*vectorization*)
                Calculate member displacements gradients
$$g_{ijk} = g_{ijk} + \frac{1}{A_i}\left(-\mathbf{v}_{imj}^T \mathbf{k}_{im} \mathbf{u}_{imk}\right) \text{ using unrolling}$$
                loop vertically
            Next member
        Next i
        {Parallel region-end}
    Next j

Step 14
Find the new design variables for loading condition k
    For design variable i = 1 until N_d (*microtasking*)
        For j = 1, N_{ac}
$$\text{sum} = \text{sum} + \left(\frac{W}{u_{jk}}\right)\frac{g_{ijk}}{\rho_i A_i \sum_{m=1}^{N_{mi}} L_{im}}$$
        Next j
        $(A_i)_{ITER+1} = (A_i)_{ITER} (\text{sum})^\zeta$
    Next design variable

Step 15
If (ITER ⟩ MAXIT)STOP
Go to step 2

8.4 APPLICATION

The parallel-vector algorithm presented in this chapter has been implemented in FORTRAN on the Cray YMP8/864 supercomputer with eight processors. The algorithm has been used for minimum weight design of several axial-load space structures and moment-resisting frames with or without diagonal bracings. The basis of design is the AISC ASD specifications (AISC, 1989). The three space truss structures are the same as those presented in Chapter 7 modeling the exterior envelope structure of the 52-, 41-, and 81-story structures and will be described here as examples 1, 2 and 3. The speedup and MFLOPS (million floating point operations per second) results of these examples will be presented. Examples 4 to 7 are steel space moment-resisting frames described in the following paragraphs. The value of 199.9×10^3 Mpa (29,000 ksi) is used for modulus of elasticity and 0.077 N/cm^3 (0.284 lb/in^3) for unit weight. The number of nodes and elements for seven example structures are summarized in Table 8.1.

Example 4 20-story space moment resisting frame.

This example is a 20-story space moment-resisting frame with a square plan and side view shown in Figure 8.1. The structure has an aspect ratio of 2.4. It has 756 nodes and 1920 members divided into 100 groups of members. A wide-flange (W) shape is selected for each group. The groups are organized

Table 8.1: Performance of the parallel-vector algorithm

Type of structure	Example	NDOF	No. of members	Aspect ratio	CPU time for vectorized code using one processor (sec.)	MFLOPS
Space truss	1	672	848	4.80	5.84	114.1
	2	2520	4100	6.80	75.02	126.1
	3	3480	5700	13.50	113.20	138.2
Space frame	4	4536	1920	2.40	149.46	130.1
	5	8856	3840	4.80	781.23	134.6
	6	13176	5760	7.20	1735.00	137.3
Space frame with bracings	7	16236	9245	8.10	2780.90	138.4

Type of structure	Example	Speedup due to parallelization on eight processors	Speedup due to vectorization only	Speedup due to vectorization & parallelization
Space truss	1	5.65	16.10	77.10
	2	6.10	17.50	89.70
	3	6.36	17.40	95.40
Space frame	4	6.10	18.10	92.70
	5	6.25	17.90	95.60
	6	6.37	17.80	97.10
Space frame with bracings	7	6.42	17.80	99.18

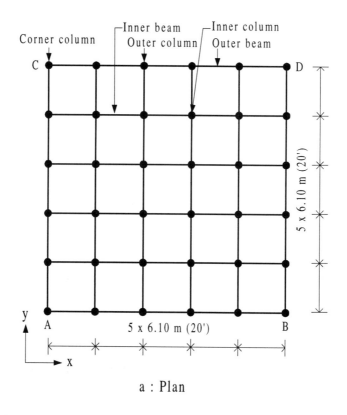

a : Plan

Figure 8.1: 20-story moment-resisting frame

as follows: Columns of each story are divided into three groups: a group of corner columns, a group of outer columns, and a group of inner columns. The beams of each floor are divided into two groups, outer beams and inner beams. The loading on the structure consists of dead load (D) of 2.78 kPa (58 psf), Live Load (L) of 2.38 kPa (50 psf), and Roof Live Load (L_r) of 2.38 kPa (50 psf). The horizontal loads in the x direction at each node on the sides AC and BD are obtained from the Uniform Building Code (UBC, 1994) using the equation $d_p = C_e C_q q_s I$

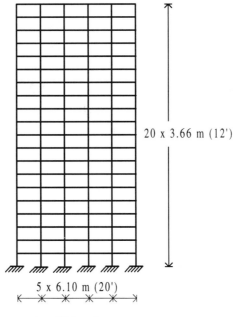

b : Side view

Figure 8.1-continued

where d_p is the design wind pressure, C_e is the combined height, exposure, and gust factor coefficient, C_q is the pressure coefficient, q_s is the wind stagnation pressure and I is the importance factor. The value of C_q for inward face is 0.8 and for the leeward face is 0.5. Assuming a basic wind speed of 70 mph (113 km/h), the value of q_s is 0.6 kPa (12.6 psf) and the importance factor is assumed to be one. The values of C_e are taken from the UBC code assuming exposure B.

The lower and upper bounds of cross-sectional areas in this example are 24.5 cm^2 (3.80 in.^{2}) and 683 cm^2 (106 in.^{2}). The maximum displacements at the top level are limited to

±29.26 cm (11.52 in.) in the direction of the wind loading (equal to 0.004 H, H being the height of the structure). The maximum number of iterations is set to 10. Figure 8.2 shows the convergence history. A minimum weight of 8.0 MN (1,809 kips) is found after 7 iterations. This translates into 0.43 kPa (9.04 lbs/ft^2) when the total weight is divided by the total floor area provided by the structure.

Example 5 40-story space moment-resisting frame

This example is a 40-story space moment-resisting frame consisting of 1476 nodes and 3840 members. The plan of this structure is the same as that of example 4 (Figure 8.1a). Its elevation is similar to that of example 4 with the same uniform story height of 3.66 m (12ft) but the total height of 146.4 m (480 ft) giving the structure an aspect ratio of 4.8. The members are divided into 200 groups of members. A W shape is selected for each group. The groups are organized similarly to example 4. The loadings on the structure and the material properties are the same as those of example 4. The lower and the upper bounds of the cross-sectional areas in this example are 24.5 cm^2 (3.80 in^2) and 1060 cm^2 (249 in^2). The displacement constraints are given as ±58.52 cm (23.04 in.) in the x direction for the nodes on the top level (equal to 0.004 H, H being the height of the structure). The maximum number of iterations is set to 10. Figure 8.3 shows the convergence history. A minimum weight of 23.9 MN (5,372 kips) is found after 7 iterations. This translates into 0.64 kPa (13.4 lbs/ft^2) when the total weight is divided by the total floor area provided by the structure.

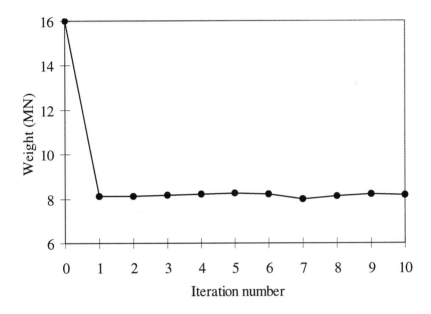

Figure 8.2: Convergence history for 20-story space moment-resisting frame

Example 6 60-story space moment-resisting frame

This example is a 60-story space moment-resisting frame consisting of 2196 nodes and 5760 members. The plan of this structure is the same as that of example 4 (Figure 8.1a). Its elevation is similar to that of example 4 with the same uniform story height of 3.66 m (12 ft) but the total height of 219.4 m (720 ft) giving the structure an aspect ratio of 7.2. Furthermore X-bracings are added covering 5-story tiers. The members are divided into 312 groups of members. A W shape is selected for each group. The groups are organized similarly to examples 4 and 5. Bracings in each five stories are assigned the same cross-sectional area. The loadings on the structure and the material

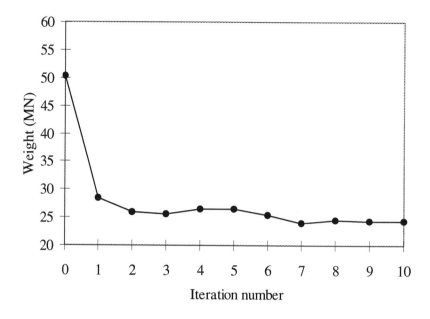

Figure 8.3: Convergence history for 40-story space moment-resisting frame

properties are the same as those of example 4. The lower and the upper bounds of the cross-sectional area are the same as those of example 4. The displacements constraints are given as ±87.78 cm (34.56 in.) in the x direction for the nodes on the top level (equal to 0.004 H, H being the height of the structure). The maximum number of iterations is set to 10. Figure 8.4 shows the convergence history. A minimum weight of 49.8 MN (11223 Kips) is found after 7 iterations. This translates into 0.89 kPa (18.70 lbs/ft^2) when the total weight is divided by the total floor area provided by the structure.

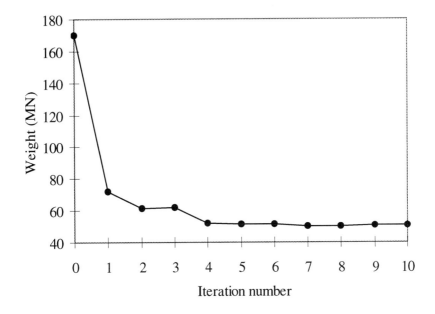

Figure 8.4: Convergence history for 60-story space moment-resisting frame

Example 7: 81-story space moment-resisting frame with cross bracings

This example is an 81-story moment-resisting frame with cross bracings. The perspective, plan views, and side view of the structure with setback are shown in Figures 8.5a to 8.5d. The structure consists of two sections: section one consisting of the lower 40 stories and section two consisting of the upper 41 stories with reduced plan. The structure has 2706 nodes, 16236 degrees of freedom, and 9245 members divided into 603 groups of members. The groups are organized as follows: columns of each story in section one are divided into four groups: a group of corner columns, a group of outer columns 1, a group of outer columns 2, and a group of inner columns, as noted in Figure 8.5.

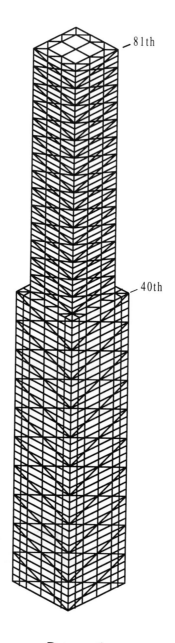

a : Perspective

Figure 8.5: 81-story space moment-resisting frame with cross bracings

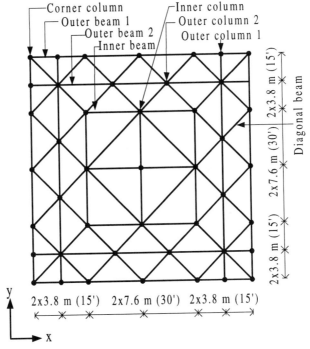

b : Plan for section 1

Figure 8.5-continued

The beams of each floor in section one are divided into four groups, outer beams 1, outer beams 2, inner beams, and diagonal beams (Figure 8.5). Columns of each story in section two are divided into three groups: a group of corner columns, a group of outer columns 2, and a group of inner columns, as noted in Figure 8.5. The beams of each floor in section two are divided into three groups: outer beams 2, inner beams, and diagonal beams (Figure 8.5). Bracings cover four stories in section 1 and three stories in section 2. Bracings in each tier are

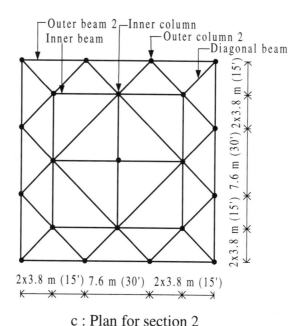

c : Plan for section 2

Figure 8.5: -continued

grouped as one. A wide-flange (W) shape is selected for each group of members.

The loading on the structure consists of dead load of 2.78 kPa (58 psf), live load of 2.38 kPa (50 psf) and roof live load of 2.38 kPa (50 psf). The horizontal loads in the x direction at each node on the inward and leeward sides are computed on the basis of the Uniform Building Code (UBC, 1994) using the equation $d_p = C_e C_q q_s I$ where d_p is the design wind pressure, C_e is the combined height, exposure, and gust factor coefficient, C_q is pressure coefficient, q_s is the wind stagnation pressure and I is the importance factor. The value of C_q for the inward face is 0.8 and for the leeward face is 0.5. Assuming a

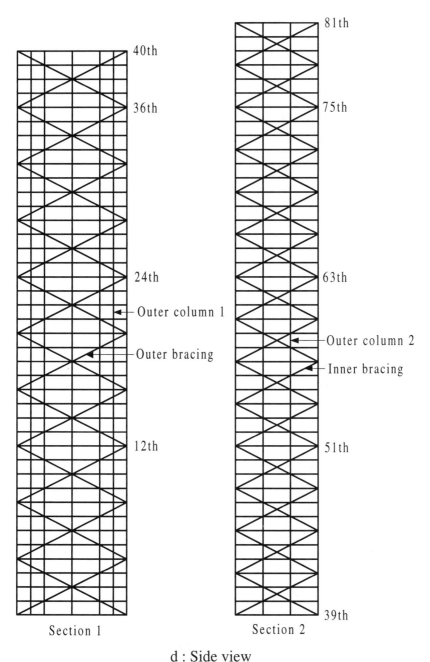

d : Side view

Figure 8.5: -continued

basic wind speed of 70 mph (113 km/h), the value of q_s is 0.6 kPa (12.6 psf) and the importance factor is assumed to be one. The values of C_c are taken from the UBC code assuming exposure B.

The lower and upper bounds of cross-sectional areas in this example are 24.5 cm^2 (380 in.^{2}) and 683 cm^2 (106 in.^{2}). The modulus of elasticity of steel is 199.9 x 10^3 Mpa (29,000 ksi) and the unit weight of material is 0.077 N/cm^3 (0.284 lb/in.^{3}). The maximum displacements at the top level are limited to 118.50 cm (46.65 in.) in the direction of the wind loading (equal to 0.004 H). Figure 8.6 shows the convergence history. A minimum weight of 103.6 MN (23316 kips) is found after 6 iterations. This translates into 1.23 kPa (25.6 lbs/ft^2) when the weight is divided by the total floor area provided by the structure. Note that the structure has a relatively high aspect (height-to-width) ratio of 8.1.

8.5 PERFORMANCE EVALUATION

The parallel-vector algorithm for optimization of structures has been implemented in Cray FORTRAN on the Cray YMP8/864 supercomputer with eight processors. The performance of the algorithm is evaluated in terms of speedup and MFLOPS. The computation time is dominated by three steps: evaluation and assembly of stiffness matrices, solution of a system simultaneous linear equation for nodal displacements and calculation of the displacement gradients. Therefore, speedup curves are presented for these steps as well as the complete optimization process. Since the program is executed in a non-dedicated multi-user environment, the parallel processing speedup is measured through a software tool called, *atexpert* (Cray, 1991).

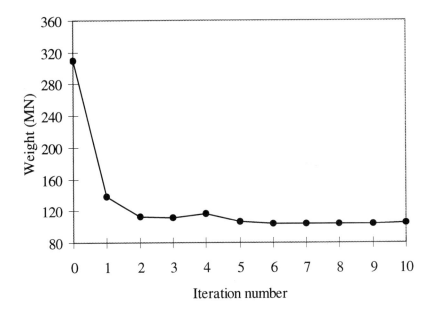

Figure 8.6: Convergence history for 81-story space moment-resisting frame

Figures 8.7 to 8.13 present the speedup results for examples 1 to 7, respectively, due to parallel processing only. Figure 8.14 shows the speedup due to both vectorization and parallel processing for space truss structures (examples 1 to 3). Similar results for space-moment resisting frames (examples 4 to 7) are presented in Figure 8.15. Performance results in terms of speedup and MFLOPS are also summarized in Table 8.1 along with the description of the size of each structure.

One clear trend can be observed in Figures 8.7 to 8.13 and Table 8.1: parallelization efficiency improves with the increase in the size of the structures. The same trend is observed

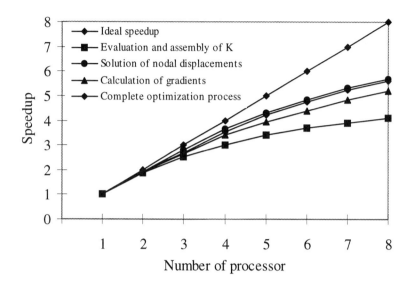

Figure 8.7: Speedups due to parallel processing for example 1

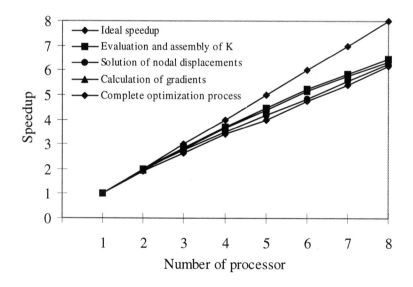

Figure 8.8: Speedups due to parallel processing for example 2

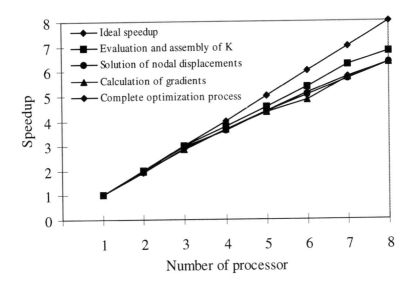

Figure 8.9: Speedups due to parallel processing for example 3

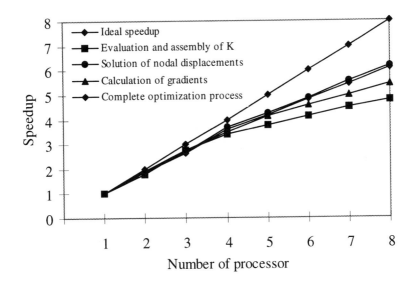

Figure 8.10: Speedups due to parallel processing for example 4

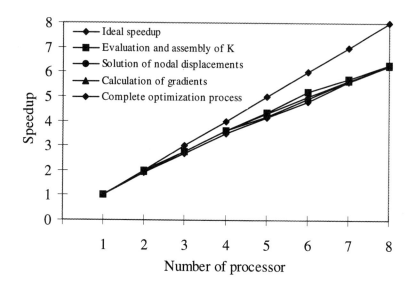

Figure 8.11: Speedups due to parallel processing for example 5

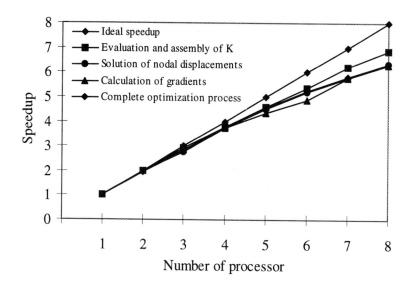

Figure 8.12: Speedups due to parallel processing for example 6

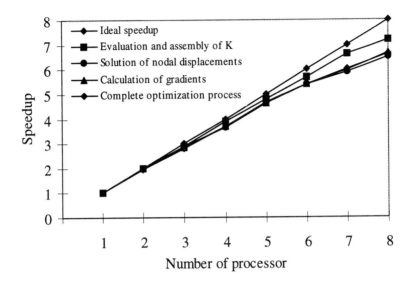

Figure 8.13: Speedups due to parallel processing for example 7

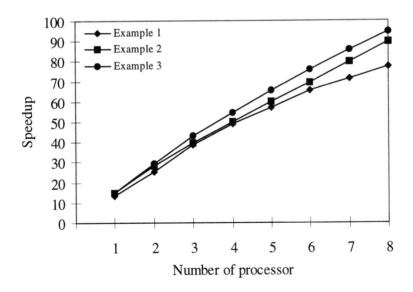

Figure 8.14: Speedups due to parallel processing and vectorization for examples 1, 2 and 3.

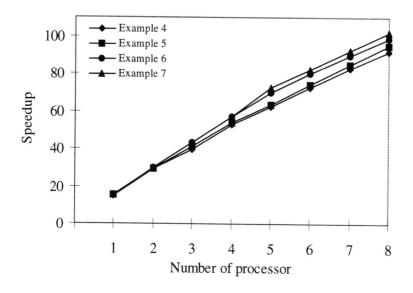

Figure 8.15: Speedups due to parallel processing and vectorization for examples 4, 5, 6 and 7.

for speedup due to vectorization but to a smaller extent. For the 81-story structure a speedup of 6.42 was achieved due to parallel processing (using eight processors) only and a speedup of 17.8 due to vectorization only. When vectorization is combined with parallel processing, a combined speedup of 99.2 is achieved. This number is somewhat smaller than the product of the previous two numbers (6.42x17.8=114.4) because parallel processing degrades vectorization to some extent. As such, vectorization and parallel processing must be combined prudently in order to achieve maximum overall speedup, especially considering the fact that speedup due to vectorization is more than the speedup due to parallel processing on a shared memory machine with a few processors.

The MFLOPS of the examples range from 133.1 to 163.6 for the vectorized code without parallelization and 114.1 to 138.7 for the code with parallelization. Without vectorization and parallelization the MFLOPS range is from 8.5 to 8.7. The overall speedup achieved due to vectorization and parallelization is very significant. It ranges from 77.1 for the smallest example (example 1) to 99.2 for the largest example (example 7).

CHAPTER 9
Optimum Load and Resistance Factor Design of Large Steel Space-Frame Structures

9.1 INTRODUCTION

In this chapter, we present a structural optimization algorithm for space moment-resisting frames subjected to actual constraints of AISC LRFD specifications (AISC, 1994). In computing the flexural strength of the members, the second order PΔ effects are considered. The algorithm is applied to minimum weight design of three space moment-resisting frames and two moment-resisting frames with diagonal bracings. Finally, the minimum weight designs based on AISC LRFD specifications are compared with the minimum weight designs based on AISC ASD specifications.

9.2 DISPLACEMENT AND STRESS CONSTRAINTS

The displacement and fabrication constraints are defined by Eqs. (7.2) and (7.3), respectively. For beam-columns, the stress constraint is obtained from the interaction of flexure and axial forces. The relationship is expressed as (AISC, 1994)

$$\begin{cases} \dfrac{P_{um}}{\phi_c P_{nm}} + \dfrac{8}{9}\left(\dfrac{M_{umx}}{\phi_b M_{nmx}} + \dfrac{M_{umy}}{\phi_b M_{nmy}}\right) \leq 1.0 & \text{for } \dfrac{P_{um}}{\phi_c P_{nm}} \geq 0.2 \\ \dfrac{P_{um}}{2\phi_c P_{nm}} + \left(\dfrac{M_{umx}}{\phi_b M_{nmx}} + \dfrac{M_{umy}}{\phi_b M_{nmy}}\right) \leq 1.0 & \text{for } \dfrac{P_{um}}{\phi_c P_{nm}} < 0.2 \end{cases} \quad (9.1)$$

where M_{umx} is the required flexural strength of the member m about the major axis, M_{umy} is the required flexural strength of member m about the minor axis, M_{nmx} is the nominal flexural strength of member m about the major axis, M_{nmy} is the nominal flexural strength of member m about the minor axis. P_{um} is the required compressive strength of member m, P_{nm} is the nominal compressive strength of member m, ϕ_b is the resistance factor for flexure (equal to 0.90), and ϕ_c is the resistance factor for compression (equal to 0.85).

The numerical value of the left-hand side of Eq. (9.1) is used as the stress scaling factor in the optimization algorithm. The nominal compressive strength of member m is computed by the following equations:

$$P_{nm} = F_{cr} A_m \qquad (9.2)$$

$$\begin{cases} F_{cr} = \left(0.658^{\lambda_c^2}\right) F_y & \text{for } \lambda_c \leq 1.5 \\ F_{cr} = \left(\dfrac{0.877}{\lambda_c^2}\right) F_y & \text{for } \lambda_c > 1.5 \end{cases} \qquad (9.3)$$

$$\lambda_c = \frac{KL_{im}}{r_y \pi} \sqrt{\frac{F_y}{E}} \tag{9.4}$$

The required flexural strength of member m is computed by the following equations:

$$\begin{cases} M_{umx} = B_{1x} M_{nmt} + B_{2x} M_{1mt} \\ M_{umy} = B_{1y} M_{nmt} + B_{2y} M_{1mt} \end{cases} \tag{9.5}$$

$$\begin{cases} B_{1x} = \dfrac{C_{mx}}{1 - P_{um}/P_{ex}} \geq 1 \\ B_{1y} = \dfrac{C_{mx}}{1 - P_{um}/P_{ey}} \geq 1 \end{cases} \tag{9.6}$$

$$\begin{cases} P_{ex} = \dfrac{\pi^2 EI_x}{(KL_{im})^2} \\ P_{ey} = \dfrac{\pi^2 EI_y}{(KL_{im})^2} \end{cases} \tag{9.7}$$

$$\begin{cases} C_{mx} = 0.6 - 0.4(M_{1mx}/M_{2mx}) \\ C_{my} = 0.6 - 0.4(M_{1my}/M_{2my}) \end{cases} \tag{9.8}$$

$$\begin{cases} B_{2x} = \dfrac{1}{1 - \sum P_u / \sum P_{ex}} \\ B_{2y} = \dfrac{1}{1 - \sum P_u / \sum P_{ey}} \end{cases} \tag{9.9}$$

where M_{nmt} is the required flexural strength in the member assuming there is no lateral translation of the frame (vertical loads only), M_{1mt} is the required flexural strength in a member as a result of lateral translation of the frame only, $\sum P_u$ is the total required axial load strength of all columns in a story, $\sum P_{ex}$ is the total computed P_{ex} about the major axis for all the columns in a story, $\sum P_{ey}$ is the total computed P_{ey} about the minor axis for all the columns in a story. The nominal flexural strengths of member m are given by long complicated equations in the AISC LRFD manual (AISC, 1994).

9.3 ALGORITHM FOR STRUCTURAL OPTIMIZATION

The parallel-vector algorithm for moment-resisting frames based on AISC ASD specifications has been discussed in Chapter 8. In the OC-based structural optimization algorithms based on the AISC LRFD specifications, most of the steps are the same as those in Chapter 8. There are two major differences between designs based on the AISC ASD and LRFD specifications. The first is the loading condition. Four nominal loads are considered in this algorithm, that is, dead load (D), live load (L), roof live load (L_r) and wind load (W). Three load combinations and the corresponding load factors are considered. They are 1.4D, 1.2D + 1.6L_r + 0.5L and 1.2D + 1.3 W + 0.5 L + 0.5L_r.

The second difference is the scaling factor for members. In the AISC LRFD specifications, the required flexural strength is determined by considering the second-order effects. This has led to the modification of several steps in the algorithm developed in Chapter 8. The algorithm is as follows:

Step 1
Read in the input data such as the initial cross-sectional properties, tolerance for displacement constraints, and the step size parameters

Step 2
Compute the member stiffness matrices
    For i = 1 until N_d
        For member m = 1 until N_{mi}
            Calculate the stiffness transformation matrix
            $T_{im}(12,12,m)$
            Calculate the member stiffness matrix
            $k_{iml}(12,12,m)$ *in local coordinates*
            Calculate the member stiffness matrix
            $k_{im}(12,12,m)$ *in global coordinates*
            $$k_{im} = T_{im}^T k_{iml} T_{im}$$
        Next member
    Next i

Assemble the structure stiffness matrix
    For member m =1 until N_m
        Assemble the member stiffness matrix into the structure stiffness matrix
$$K = \sum_{m=1}^{N_m} k_{im}(12,12,m)$$
    Next member

Modify the structure stiffness matrix to include the boundary conditions
    For nodal degree of freedom j = 1 until N
        Modify the structure stiffness matrix to include the boundary conditions
    Next j

Step 3
Assemble the loads
    For loading condition k = 1 until L
        For nodal degree of freedom j = 1 until N

228 Optimum LRFD of Large Steel Space-Frame Structures

$$\mathbf{F}_k = \sum_{j=1}^{N_m} \mathbf{f}_{jk}$$

    Next j
Next loading

Step 4
Solve the nodal displacements due to vertical loads only
    For loading condition k = 1 until L
        Solve for global nodal displacements $(\mathbf{u}_{vk})$
$$\mathbf{K}\mathbf{u}_{vk} = \mathbf{F}_{vk}$$
    Next loading

Step 5
Find the member forces due to vertical loads only
    For loading condition k = 1 until L
        For i = 1 until N_d
            For member m = 1 until N_{mi}
                Calculate the member displacement vector in local coordinates
$$\mathbf{u}_{imvk\,l} = \mathbf{T}_{im}^T \mathbf{u}_{imv\,k} \mathbf{T}_{im}$$
                Calculate the member forces
$$\mathbf{F}_{imvk\,l} = \mathbf{k}_{im\,l} \mathbf{u}_{imvk\,l}$$
                Add the member fixed-end moments
$$\mathbf{F}_{imvk\,l} = \mathbf{F}_{imvk\,l} + \mathbf{F}_{imk}$$
            Next member
        Next i
    Next loading

Step 6
Solve the nodal displacements due to horizontal loads only
    For loading condition k = 1 until L
        Solve for global displacements $(\mathbf{u}_{hk})$

$$\mathbf{Ku}_{hk} = \mathbf{F}_{hk}$$
Next loading

Step 7
Find the member forces due to horizontal loads only
    For loading condition k = 1 until L
        For i = 1 until N_d
            For member m = 1 until N_{mi}
                Calculate the member displacement vector in local coordinates
$$\mathbf{u}_{imhk\,l} = \mathbf{T}_{im}^T \mathbf{u}_{imhk} \mathbf{T}_{im}$$
                Calculate the member forces
$$\mathbf{F}_{imhk\,l} = \mathbf{k}_{im\,l}\mathbf{u}_{imhk\,l}$$
            Next member
        Next i
    Next loading

Step 8
Find the displacement vector due to vertical and horizontal loads
    For loading condition k = 1 until L
        For nodal degree of freedom j = 1 until N
            Calculate the member displacement vector in local coordinates
$$\mathbf{u}_k = \mathbf{u}_{vk} + \mathbf{u}_{hk}$$
        Next nodal degree of freedom
    Next loading

Step 9
Find the maximum displacement scaling factor
    For loading condition k = 1 until L
        For nodal degree of freedom j = 1 until N

Find the maximum displacement scaling factor $(SFD)_k$

$$(SFD)_k = \max \frac{u_{jk}}{r_j}$$

Next j
Next loading

Step 10
Find the stress scaling factor for members
For loading condition k = 1 until L
For beam member m = 1 until N_{be}
Find the value of interaction from Eq. (9.1) for beam and use it as the stress scaling factor
Next beam member
For beam member m = $N_{be} + 1$ until N_m
Find the value of interaction from Eq. (9.1) for column and use it as the stress scaling factor
Next column member
Next loading

Step 11
Find the maximum stress scaling factor
For loading condition k = 1 until L
For member m = 1 until N_m
Find the maximum stress scaling factor $(SFS)_k$
$(SFS)_k = \max SF_{mk}$
Next member
Next loading

Step 12
Scale the moments of inertia and select a W shape
If $|SFD - 1.0\rangle$ tol. and $N_{an} = 1$

For loading condition k
    For member m = 1 until N_d
        $I_{mx} = I_{mx} * (SFD)_k * \gamma_1$
    Next member
    For member m = 1 until N_d
        Select W shapes using the mapping strategy described in Chapter 7
    Next member
    $N_{an} = N_{an} + 1$
Endif
Go to step 2
    If $|SFD - 1.0| >$ tol. and $N_{an} = 2$
    For loading condition k
        For member m = $N_d / 2$ until N_d
            $I_{mx} = I_{mx} * (SFD)_k * \gamma_2$
        Next member
        For member m = $N_d / 2$ until N_d
            Select W shapes using the mapping strategy described in Chapter 7
        Next member
        $N_{an} = N_{an} + 1$
    Endif
Go to step 2
    If $|SFD - 1.0| >$ tol. and $N_{an} = 3$
    For loading condition k
        For member m = $3 * N_d / 4$ until N_d
            $I_{mx} = I_{mx} * (SFD)_k * \gamma_3$
        Next member
        For member m = $3 * N_d / 4$ until N_d
            Select W shapes using the mapping strategy described in Chapter 7

Next member
$N_{an} = N_{an} + 1$
Endif
Go to step 2
If $|SFD - 1.0| >$ tol. and $3 < N_{an} \leq 5$
Find those members whose stress constraints are violated
For $i = 1$ until N_d
$N_c = 0$
For member $m = 1$ until N_{mi}
If $SF_{mk} > SF$
$SF = SF_{mk}$
$N_c = N_c + 1$
$im = i$
Endif
Next member
If $(N_c > 1)$ then
$I_{imx} = I_{imx} * SF * 1.1$
Endif
Select W shapes using the mapping strategy described in Chapter 7
Next i
$N_{an} = N_{an} + 1$
Endif
Go to step 2

Step 13

Calculate the weight of the structure
For member $m = 1$ until M
$$W = \sum_{m=1}^{M} A_m * L_m * \rho_m$$
Next member

Step 14
Check the weight of the structure
   If $\left(W^{new} \rangle W^{old} \text{ and } \zeta \rangle 0.1\right)$ $\zeta = 0.5 * \zeta$ and go to step 18
   ITER = ITER +1

Step 15
Find the active displacement constraints and their corresponding Lagrange multipliers λ_{jk}

Step 16
Calculate the virtual nodal displacements
   For active displacements a = 1 until N_{ac}
      Find the virtual nodal displacements due to a unit load in the direction of the jth active displacement constraint
      $\mathbf{Kv}_{ja} = \mathbf{F}_{ja}$
   Next active constraint
   For j = 1 until N_{ac}
      For i = 1 until N_d
         For member m = 1 until N_{mi}
            Find the member virtual nodal displacements $\mathbf{v}_{imj}$
         Next member
      Next i
   Next j

Step 17
Calculate displacement gradients
   For i = 1 until N_d
      For member m = 1 until N_{mi}
         Calculate member stiffness matrix $\mathbf{k}_{im}(12,12,m)$
   Next member
   For loading condition k
      For i = 1 until N_d

For member m = 1 until N_{mi}
    Calculate member displacements vector $\mathbf{u}_{imk}$
Next member
Next i
For loading condition k
For j = 1, N_{ac}
  For i = 1 until N_d
    For member m = 1 until N_{mi}
    Calculate member displacement gradients
$$g_{ijk} = g_{ijk} + \frac{1}{A_i}\left(-\mathbf{v}_{imj}^T \mathbf{k}_{im} \mathbf{u}_{imk}\right)$$
    Next member
  Next i
Next j

Step 18
Find the new design variables for loading condition k
For design variable i = 1 until N_d
  For j = 1, N_{ac}
$$\text{sum} = \text{sum} + \left(\frac{W}{u_{jk}}\right)\frac{g_{ijk}}{\rho_i A_i \sum_{m=1}^{N_{mi}} L_{im}}$$
  Next j
$$(A_i)_{ITER+1} = (A_i)_{ITER} (\text{sum})^\zeta$$
Next design variable

Step 19
If $(\text{ITER} \rangle \text{MAXIT})$ STOP
Go to step 2

9.4 APPLICATION

The algorithm has been applied to the minimum weight design of four steel space-frame high-rise building structures. Four different types of loads are considered: dead load (D), live load (L), roof live load (L_r), and wind load (W). Three load combinations are considered, per AISC LRFD specifications: 1.4D, 1.2D + 1.6L_r +0.5L, and 1.2D + 1.3W + 0.5L + 0.5 L_r. The first two examples are space moment-resisting frames. The last two examples are moment-resisting frames with cross bracings. The material is steel with modulus of elasticity of 199.9 x 10^3 Mpa (29,000 ksi) and the unit weight of the material of 0.077 N/cm^3 (0.284 lb/in^3). For designs based on the AISC ASD specifications, the maximum drift is limited to 0.004H (H is the total height of the structure). For designs based on the LRFD specification, we increase this value by 30% to 0.0052H to reflect the coefficient 1.3 in the aforementioned LRFD wind load combination.

Example 1: 20-story space moment-resisting frame

This example is a 20-story space moment-resisting frame with a square plan as shown in Figures 8.1a and 8.1b. The configuration and loads are the same as those of example 4 of Chapter 8. The displacement constraints are given as ±38.05 cm (14.98 in.) in the x direction for nodes on the top level (equal to 0.0052H). The maximum number of iterations in this example is set to 10. Figure 9.1 shows the convergence history. A minimum weight of 7.45 MN (1,675 kips) is found after eight iterations. This translates into 0.40 kPa (8.4 lb/ft^2) when the total weight is divided by the total floor area provided by the structure.

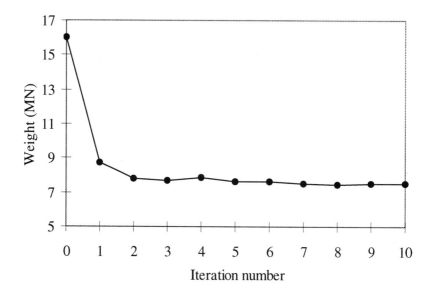

Figure 9.1: Convergence history for 20-story space moment-resisting frame

Example 2: 40-story space moment-resisting frame

This example is a 40-story space moment-resisting frame consisting of 1476 nodes and 3840 members. The configuration and loads are the same as those of example 5 in Chapter 8. The displacement constraints are given as ±76.10 cm (29.96 in.) in the x direction for nodes on the top level (equal to 0.0052H). The maximum number of iterations in this example is set to 10. Figure 9.2 shows the convergence history. A minimum weight of 22.38 MN (5,032 kips) is found after seven iterations. This translates into 0.60 kPa (12.6 lb/ft^2) when the total weight is divided by the total floor area provided by the structure.

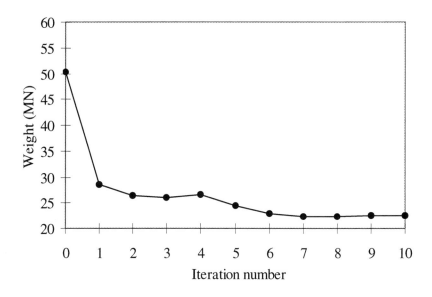

Figure 9.2: Convergence history for 40-story space moment-resisting frame

Example 3: 60-story space moment-resisting frame

This example is a 60-story space moment-resisting frame consisting of 2196 nodes and 5856 members. The configuration and loads are the same as those of example 6 in Chapter 8. The displacement constraints are given as ±114.11 cm (44.92 in.) in the x direction for nodes on the top level (equal to 0.0052H). Figure 9.3 shows the convergence history. A minimum weight of 46.4 MN (10,436 kips) is found after nine iterations. This translates into 0.83 kPa (17.4 lb/ft^2) when the total weight is divided by the total floor area provided by the structure.

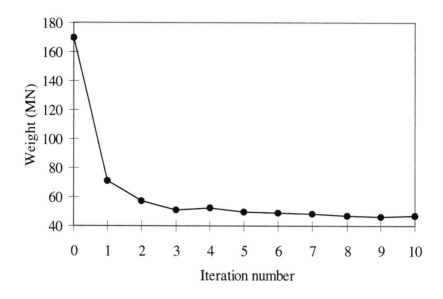

Figure 9.3: Convergence history for 60-story space moment-resisting frame

Example 4: 81-story space moment-resisting frame with cross bracings

The 81-story space moment-resisting frame with cross bracings consists of 2706 nodes and 9245 members. The configuration and loads are the same as those of example 6 in Chapter 8. The displacement constraints are given as ±154.05 cm (60.65 in.) in the x direction for nodes on the top level (equal to 0.0052H). Figure 9.4 shows the convergence history. A minimum weight of 96.79 MN (21,788 kips) is found after seven iterations. This translates into 1.15 kPa (23.98 lb/ft^2). when the total weight is divided by the total floor area provided by the structure.

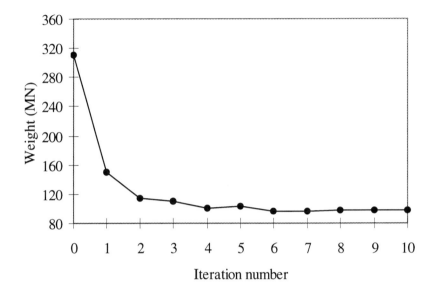

Figure 9.4: Convergence history for 81-story space moment-resisting frame with bracings

9.5 COMPARISON OF DESIGNS BASED ON ASD AND LRFD SPECIFICATIONS

The multi-constraint discrete algorithm for optimization of steel moment-resisting space frames has been implemented in FORTRAN on the Cray YMP8/864 supercomputer. The code has been vectorized using strategies presented in Chapter 6. The optimization convergence histories presented in Figures 9.1 to 9.4 show consistently good performance for large structures.

A comparison of minimum weight designs based on AISC LRFD and ASD specifications for the four examples presented in this chapter is presented in Figure 9.5. This figure shows that in all examples the minimum weight design based on the LRFD code resulted in a lighter structure. The saving in the weight is in the range of 6.3% to 7.4%. In this work, the

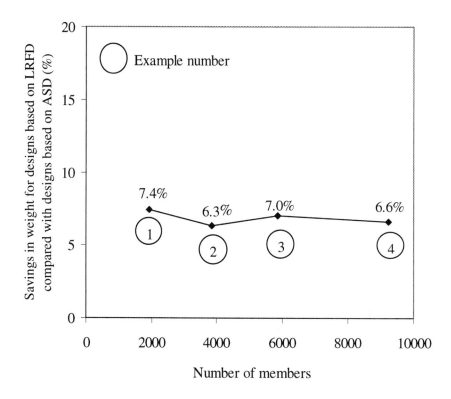

Figure 9.5: Savings in weight designs based on LRFD specifications compared with designs based on ASD specifications as a function of the number of members

torsional effect has been neglected. To include this effect, additional constraints resulting from the twisting of the structure have to be incorporated in the algorithm.

REFERENCES

Adeli, H. (1992a), *Supercomputing in Engineering Analysis, Marcel Dekker*, New York, NY.

Adeli, H. (1992b), *Parallel Processing in Computational Mechanics*, Marcel Dekker, New York, NY.

Adeli, H., Ed. (1994), *Advances in Design Optimization*, Chapman and Hall, London.

Adeli, H. and Al-Rijleh, M.M. (1987), "A Knowledge Based Expert System for Design of Roof Trusses", Microcomputers in Civil Engineering, Vol. 2, No. 1, pp. 179-195.

Adeli, H. and Balasubramanyam, K.V. (1988), *Expert System for Structural Design*, Prentice-Hall, Engelwood-Cliffs, NJ.

Adeli, H. and Cheng, N.T. (1993), "Integrated Genetic Algorithm for Optimization of Space Structures", Journal of Aerospace Engineering, ASCE, Vol. 6, No.4, pp. 315-328.

Adeli, H. and Cheng, N.T. (1994), "Concurrent Genetic Algorithms for Optimization of Large Structures", Journal of Aerospace Engineering, ASCE, Vol. 7, No.3, pp. 276-296.

Adeli, H. and Kamal, O. (1986), "Efficient Optimization of Space Trusses, Computers & Structures", Vol. 24., No. 3, pp. 501-511.

Adeli, H. and Kamal, O. (1992a), "Concurrent Optimization of Large Structures Part I-Algorithm", Journal of Aerospace Engineering, ASCE, Vol. 5, No. 1, pp. 79-90.

Adeli, H. and Kamal, O. (1992b), "Concurrent Optimization of Large Structures Part II-Applications", Journal of Aerospace Engineering, ASCE, Vol. 5, No. 1, pp. 91-110.

Adeli, H. and Kamal, O. (1993), *Parallel Processing in Structural Engineering*, Elsevier Applied Science, London.

Adeli, H., Kamat, M.P., Kulkarni, G., and Vanluchene, R.D. (1993), "High-Performance Computing in Structural Mechanics and Engineering", Journal of Aerospace Engineering, ASCE, Vol. 6, No.3, pp. 249-267.

Adeli, H. and Saleh, A. (1997), "Optimal Control of Adaptive/Smart Bridge Structures", Journal of Structural Engineering, ASCE, Vol. 123, No.2, pp. 218-226.

Adeli, H. and Vishnubhotla, P. (1992), "Parallel Machines", in Adeli, H., Ed., *Parallel Processing in Computational Mechanics*, Marcel Dekker, New York, pp. 1-20.

AISC (1989), *Manual of Steel Construction-Allowable Stress Design*, American Institute of Steel Construction, Chicago, Illinois.

AISC (1994), *Manual of Steel Construction-Load and Resistance Factor Design*, American Institute of Steel Construction, Chicago, Illinois.

Anonymous (1978), *European Convention for Constructional Steel Work*, European Convention for Construction Steel Work.

Agarwal, T.K., Storaasli, O.O., and Nguyen, D.T. (1990), "A Parallel-Vector Algorithm for Rapid Structural Analysis on High Performance Computers", Proceedings of the AIAA/ASME/ASCE/AHS 31-st SDM Conference, Long Beach, CA.

Chan, C.M. 1992, "An Optimality Criteria Algorithm for Tall Steel Building Design Using Commercial Standard Sections", Journal of Structural Optimization, Vol. 5, pp. 26-29.

Cray, (1990), CF77 *Compiling System*, Vol. 1-3., Cray Research Inc., Mendota Heights, Minnesota.

Cray, (1991), CF77 *Compiling System*, Vol. 4., Cray Research Inc., Mendota Heights, Minnesota.

Dumonteil, P. (1992), "Simple Equations for Effective Length Factors", Engineering Journal, AISC, Third Quarter, pp. 111-115.

Encore (1985), *Multimax Technical Summary*, Encore Computer Corporation, Marlboro, MA.

Encore (1988), *Encore Parallel Threads Manual*, Encore Computer Corporation, Marlboro, MA.

Fletcher, R. and Reeves, C.M. (1964), "Function Minimization by Conjugate Gradients", Computer Journal, Vol. 7, pp. 149-154).

Grierson, D.E. and Chan C.M. (1993), "An Optimality Method for Tall Steel Buildings", Advances in Engineering Software, Vol. 16, No. 2, pp. 119-125.

Golub, H.G. and Van Loan, C.F. (1991), *Matrix Computations*, The John Hopkins University Press, London.

Hsu, H.L. and Adeli, H. (1991), "Microtasking Algorithm for Optimization of Structures", International Journal of Supercomputer Applications, Vol. 5, No. 2., pp. 81-90.

John, K.V. and Ramakrishnan C.V. (1990), "Discrete Optimal Design of Trusses with Stresses and Frequency Constraints", Journal of Engineering Computing, Vol. 7, pp. 37-47.

Khot, N.S. and Berke, L. (1984), "Structural Optimization Using Optimality Criteria Methods", Atrek, E., Gallagher, R.H., Ragsdell K.M., and Zienkiewicz O.C., Ed., New Directions in Optimum Structural Design, John Wiley, New York, pp. 47-74.

Luenberger, D. G. (1984), *Introduction to Linear and Nonlinear Programming*, Addison-Wesley, Reading, MA.

Ortega, J.M. (1989), *Introduction to Parallel and Vector Solution of Linear Systems*, Plenem Press, New York.

Poole, E.L. and Overman, A.L. (1998), "The Solution of Linear System of Equations with a Structural Analysis Code on the Cray-2", NASA CR-4159.

Saleh, A. and Adeli, H. (1994), Microtasking, Macrotasking, and Autotasking for Structural Optimization", Journal of Aerospace Engineering, ASCE Vol. 7, No. 2, pp. 156-174.

Shivakumar, K.N., Bigelow, C.A., and Newman, Jr., J.C. (1992), "Parallel Computation in a Three-dimensional Elastic-plastic Finite-element Analysis", Computers & Structures Vol. 43, No. 2, pp. 237-245.

Soegiarso, R. and Adeli, H. (1994), "Impact of Vectorization on Large-scale Structural Optimization", Journal of Structural Optimization Vol. 7, No. 112, pp. 117-125.

Soegiraso, R. and Adeli, H. (1995), "Parallel-Vector Algorithms for Analysis of Large Space Structures", Journal of Aerospace Engineering, ASCE, Vol. 8, No. 1 pp. 54-67.

Soegiraso, R. and Adeli, H. (1996), Optimization of Large Steel Structures Using Standard Cross Sections", Engineering Journal, AISC, Vol. 33, No.3, pp. 83-117.

Soegiarso, R. and Adeli, H. (1997a), "Optimum Load and Resistance Factor Design of Steel Space Frame Structures", Journal of Structural Engineering, ASCE Vol. 123, No. 2, pp. 184-192.

Soegiarso, R. and Adeli, H. (1997b), "Optimization of Large Space Frame Steel Structures", Engineering Journal, AISC, Vol. 34, No. 2, pp. 54-60.

Soegiarso, R. and Adeli, H. (1998), "Parallel-Vector Algorithm for Optimization of Large Steel Structures on a Shared-Memory Machine", Computer-Aided Civil and Infrastructure Engineering, Vol. 13, No.3, pp. 207-217.

Storaasli, O.O., Nguyen, D.T. and Agarwal, T.K. (1990), "Parallel-vector Solution of Large-scale Structural Analysis Problems on Supercomputers", AAIA Journal, Vol. 28, No. 7, pp. 1211-1216.

UBC (1994), *Uniform Building Code*, International Conference of Building Officials, Whittier, California.

Vishnubhotla, P. and Adeli, H. (1992), "Parallel Programming Language and Techniques", in Adeli, H., Ed., *Parallel Processing in Computational Mechanics*, Marcel Dekker, New York, pp. 21-32.

Venkayya, V.B., (1971), "Design of Optimum Structures", Journal of Computers and Structures, Vol. 1, pp. 265-309.

Zima, H. and Chapman, B. (1991), *Supercompilers for Parallel and Vector Computers*, ACM Press, New York.

SUBJECT INDEX

Active constraints, 117, 125
Address registers, 74
AISC, 139, 140, 142, 145, 148, 157, 162
Analytical approach, 113
Anti-dependency, 80
ASD, 140, 142, 145, 162
Atexpert, 101
Autotasking, 86
Axial force, 17, 18, 28
Axial-load truss, 1

Backward solution, 55, 57, 60
Bandwidth, 65
Bank conflict, 90
Bank-cycle, 83
Barriers, 79
Beam theory, 30, 32, 34, 36
Bending-moment, 18, 26
Bernoulli-Euler, 20, 21
Buckling, 139, 140, 141, 142, 148

Buckling constraints, 180

Cholesky decomposition, 53, 59, 60, 71
Column sweep, 95
Conjugate gradient, 63
Control dependency, 80
Convergence tolerance, 62, 63, 65
Cray YMP, 79

Data stream, 89, 95
Dependency, 79, 80, 81, 83
Design linking, 140, 141
Diagonal preconditioning, 65
Direct methods, 53, 65
Discrete optimization, 139, 140, 146, 148
Displacement constraints, 117
Displacement method, 5

Effective length, 181, 182

Eigenvalues, 62, 64

Feasible domain, 112
Finite difference, 113
Fletcher and Reeves, 62
Floating point operations, 101
Flow dependency, 80
Force method, 5
Forward solution, 55, 57, 60
Functional unit, 89, 90
Fundamental period, 74

Global coordinate, 6, 8, 17, 28
Global displacements, 9, 14, 24
Global stiffness, 66, 68
Gradient, 60, 61, 62, 63, 64
Guard and Endgurad, 87
Guarded region, 186

High-performance, 1
Homogeneous, 6
Hooke's law, 6, 18, 28

Inactive constraints, 117
Index register, 74
Indirect methods, 86
Interval halving, 149
Integer mapping, 140
Integer search, 140, 149, 151, 157
Iterative, 53, 61

Jacobi preconditioning, 64

Kuhn-Tucker, 118

Lagrange multipliers, 117, 120, 125
Lateral drift, 184
Linear equations, 53, 54, 60, 61, 62, 66

Linear programming, 112
Local coordinate, 6, 8
Loop splitting, 83
Loop unrolling, 83, 89, 93, 93, 95
LRFD 140, 142, 145,
LU decomposition, 53, 54, 56, 60, 70

Macrotasking, 86
Master processor, 86
Master task, 86
Member forces, 11, 14, 28
Member stiffness, 6, 8, 12, 16, 23, 66, 67, 68, 69
Memory-to-memory, 77
Microtasking, 86, 90, 92, 93, 98
Modulus of elasticity, 105, 106
Moment-resisting frame, 1
Multi-constraint, 3
Multitasking, 86, 95

Objective function, 61, 62
Optimality criteria, 112, 114, 139, 140, 146, 147
Output dependency, 80
Overhead, 74, 77, 83, 99

Parallel algorithms, 1
Parallel processing, 2
Parallel region, 86
Parallelization, 3
Parallel-vector, 73, 179, 180, 185, 202, 203
Partitioned, 13
Pipelining, 77, 78
Positive definite matrix, 57, 60, 65, 66
Preconditioned conjugate gradient, 53, 63
Private memory, 74
Quadratic programming, 112

Racing conditions, 87
Recurrence equation, 144
Registers, 74, 77, 79

Scalar register, 74
Scaling factor, 113, 123
Search vector, 61, 63
Sensitivity analysis, 113
Shared memory, 73, 74
Shear force, 17, 18, 28
Slave processors, 86, 87
Slenderness ratio, 142
Space frame, 38, 183, 186, 189, 191, 193, 200
Space truss, 183, 187, 190, 192, 197, 200, 202
Speedup, 73, 77
Steepest descent method, 60, 61, 62, 63, 112, 113
Step size, 112, 120, 121, 122
Stiffness coefficients, 6, 7, 18, 20, 21, 22, 28
Stiffness method, 5

Stride, 90, 95
Strongly connected components, 81
Subroutine inlining, 83
Supercomputer, 1
Synchronization, 79

Transformation, 9, 10
Tree network, 140, 149, 150, 151, 155, 157

Unbraced frames, 181, 182, 184
Uniform Building Code, 101, 103

Vector chaining, 79
Vector register, 77, 79
Vectorization, 2, 3
Virtual load, 114, 115, 116, 117, 128

Wall-clock time, 97, 101